W9-BST-628

AN ILLUSTRATED GUIDE TO
MODERN
ATTACK
AIRCRAFT

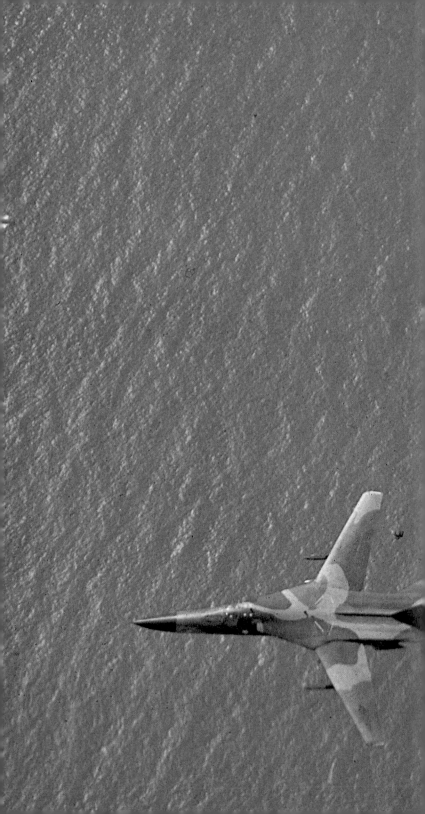

AN ILLUSTRATED GUIDE TO
MODERN
ATTACK
AIRCRAFT

Mike Spick

An Arco Military Book
Prentice Hall Press • New York

A Salamander Book

Copyright © 1987 by Salamander Books
Ltd.

All rights reserved, including the right of
reproduction in whole or in part in any
form

An Arco Military Book

Published in 1987 by Prentice Hall Press
A Division of Simon & Schuster, Inc.
Gulf + Western Building
One Gulf + Western Plaza
New York, NY 10023

PRENTICE HALL PRESS is a trademark
of Simon & Schuster, Inc.

Originally published by
Salamander Books Ltd., London

This book may not be sold outside the
United States of America and Canada.

**Library of Congress
Cataloging-in-Publication Data**

Spick, Mike.
  An illustrated guide to modern attack aircraft.

  (An Arco Military Book)
  1. Attack planes. I. Title. II. Series.
UG1242.A28S65   1987   358.4'3   86-25572
ISBN 0-13-451097-6

10 9 8 7 6 5 4 3 2 1

First Prentice Hall Press Edition

All correspondence concerning the content
of this book should be addressed to
Salamander Books Ltd.,
52 Bedford Row,
London WC1R 4LR,
United Kingdom.

Contents

Credits

Author: Mike Spick is the author of several works on modern combat aircraft and the tactics of air warfare, including Salamander's *Modern Air Combat* and *Modern Fighting Helicopters* (both with Bill Gunston) and Fact Files on the F-4 Phantom (with Doug Richardson), F-14 Tomcat and F/A-18 Hornet, as well as a companion volume to this work, *An Illustrated Guide to Modern Fighter Combat*.

Editor: Bernard Fitzsimons
Art Editor: Mark Holt
Designed by TIGA

Diagrams: TIGA

Typeset: The Old Mill, London
Colour reproduction by Melbourne Graphics
Printed in Belgium by Proost International Book Production, Turnhout

Acknowledgements: The publishers are grateful to all the companies and other organisations who supplied photographs for use in this book.

Weapon Systems

THE definition of a modern attack aircraft adopted here is a fast jet designed or adapted for the specific task of delivering air-to-surface ordnance in a tactical situation against a modern air defence system. It is necessary to adopt this definition because almost every military aircraft other than large transports has some air-to-ground capability.

Most interceptors and air superiority fighters have air-to-surface as a secondary role, as has every armed trainer and counter-insurgency aircraft, while bombers such as the B-52 and B-1B can carry a simply enormous amount of ordnance.

It has therefore been necessary to be selective. The multi-role fighters have been covered in a companion volume; 'tactical' rules out the strategic bombers, although there can be no guarantee that they would not be called upon to intervene in a tactical situation if it were desperate enough; while 'fast jet' eliminates the largely propeller-driven counter-insurgency types, such as the Socata Guerrier. Finally, the term 'against a modern air defence system' is used as a standard against which to assess and eliminate many of the less capable armed trainers. The complexity of the subject makes it impossible to be totally consistent: while the accent is on purpose-designed attack aircraft, a few multi-role fighters have crept in, notably the F-15, due to the adaptation of the F-15E strike fighter; the F-16, which some air arms use primarily in the attack role; and the F/A-18, which was optimised for the dual role and does not have a primary role.

The term 'attack' refers primarily to overland operations, although certain anti-shipping missions also have to be included. A wide spectrum of aircraft has therefore had to be covered, ranging from the barely affordable to, at the bottom end of the market, the barely credible, taking in such specialised products as the tank-busting A-10 and close air support Su-25 Frogfoot on the way. A degree of specialisation is inevitable, depending on the perceived threat and the available funding. Aircraft procurement for differing air arms

Right: A B-52 strikes a Viet Cong target in South Vietnam. Area bombing had little effect.

Below: Even light aircraft can carry weapons, as this Socata Guerrier shows, but it does not rate as an attack aircraft.

varies from 'we need X aircraft in order to mount Y sorties but can only afford $Z million', to 'we must have an aircraft optimised to fly this specific mission, and another to fly that one'. As no air arm has an unlimited budget, both end up with a compromise, but in the first case perhaps only an armed advanced trainer will fit, with a marginal capability to do the job, while in the second case two first-class aircraft are developed to meet the specific requirements, although perhaps in smaller numbers than were originally envisaged.

A third view is typified by the Royal Air Force which, when formulating the requirement that led to the Tornado GR.1, argued that numerical strength/cost was basically irrelevant, and that the nub of the matter was how many targets could be attacked accurately, within a given time span, for a given attrition rate, and presumably also against a given level of defence strength. This nononsense approach involved concentrating on what was perceived as the major threat, the negation of which would reduce the threat levels in other areas, reducing both the

menace and the effort required to deal with it.

Of course, in practice, the unexpected becomes the norm. A truism of all warfare is that one rarely fights the war for which one has been equipped and trained, but is forced to manage with the available equipment in unforeseen circumstances. It would therefore be unwise to have a totally specialised aircraft that could fly just one mission in the tactical arena.

Instead of specialised aircraft, the modern trend is to go for specialised weapons. The bottom line in attack work is ordnance on target. The destruction of various types of target calls for weapons designed to do that very job; the function of the aircraft then becomes the penetration of the defences to a point from where the target can be identified and the weapons launched. The techniques of penetration are dealt with later; here we need to know something of the weapons themselves, but before we examine them in detail, we first need to make a few generalisations.

Firstly, there are many types of air-to-ground weapons, normally optimised to destroy different types of

Above: The maximum load of air-to-ground weapons carried by a single F-111 is amazing, though a combat load would be much less. The groupings show the total load for each type. No smart or stand-off weapons are shown—the F-111 was supposed to make them unnecesary.

Below: The small F-16 is able to carry an impressive load with many different types of store. Tactical nukes make up the front row, and laser-guided Paveways can be seen among the iron bombs—an example of the multi-role fighter's need for specialised weapons.

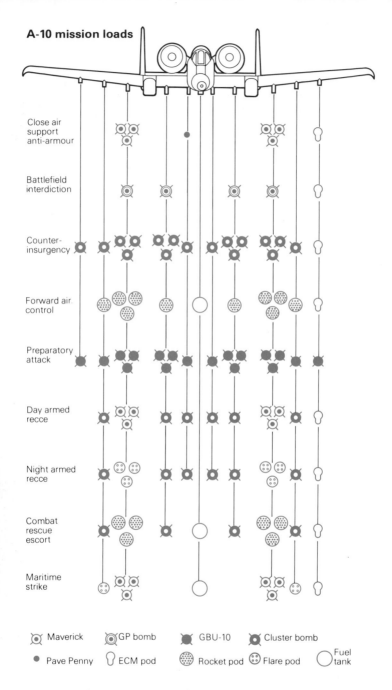

A-10 mission loads

Close air support anti-armour		GP bomb (×3)		Pave Penny			GP bomb (×3)		ECM pod
Battlefield interdiction		Maverick	Maverick		Maverick	Maverick			ECM pod
Counter-insurgency	Cluster	Cluster Maverick ×2	Maverick ×2	Cluster	Cluster	Maverick ×2	Cluster		ECM pod
Forward air control		Rocket pods	Rocket pod	Fuel tank	Rocket pod	Rocket pods			ECM pod
Preparatory attack	GBU-10	GBU-10 GBU-10 ×2	GBU-10 ×2		GBU-10 ×2	GBU-10	GBU-10		
Day armed recce		Cluster GP bomb ×2	Cluster	Cluster	Cluster	GP bomb ×2	Cluster		ECM pod
Night armed recce		Cluster Flare pods	Cluster	Cluster	Cluster	Flare pods	Cluster		ECM pod
Combat rescue escort		Cluster Rocket pods	Cluster	Fuel tank	Cluster	Rocket pods			ECM pod
Maritime strike		Maverick ×3		Fuel tank		Maverick ×2 Flare pod			ECM pod

Legend:

- ◎ Maverick
- ◎ GP bomb
- ■ GBU-10
- ◎ Cluster bomb
- ● Pave Penny
- ∩ ECM pod
- ⊛ Rocket pod
- ⊙ Flare pod
- ○ Fuel tank

Above: The type of weapon is determined by the mission, and most aircraft are cleared to carry different types. In practice, logistic problems limit availability.

Right: Harriers and Jaguars are both used for interdiction, the Harriers with their quick reaction time hitting targets just behind the front while the Jaguars penetrate deeper.

14

target, but often simply variants on a theme. Standardisation would greatly assist procurement, storage, operational training and costs, but there seems little chance of achieving it, despite the advantages. Additionally, while many types of aircraft are cleared for the carriage and release of many types of stores, it is difficult to build up a worthwhile selection at any one airfield, as the range is too vast. Units consequently tend to specialise, according to their projected role and their weapons stockpile, which is also based on the role.

A further consideration is the type of aircraft flown and its capabilities, which in the attack role means payload/range, the ability to locate a given target and survivability. These are qualified by the airfield's distance from the front line or the target, and possibly its ability to operate from a damaged runway — in other words, its short field performance.

Taking a 'worst case' conventional war in Central Europe between NATO and WARPAC forces, specialisation would lead to American A-10s roaming the FLOT at low level,

possibly backed by Luftwaffe Alpha Jets; British Harriers on short interdiction missions to disrupt reinforcements ten miles or so behind the lines, before they had time to deploy, paying special attention to choke points such as bridges and roads through constricted areas such as gorges and using their close-up basing capability to give rapid reaction and quick turn-around; British Jaguars and Belgian F-16s going deeper to strike at supply lines and depots, railways and marshalling yards; and on deeper penetration missions still, British and German Tornados and American F-111s going after airfields, command centres, and hardened targets which demanded absolute precision, preferably at night or in weather conditions that would hamper the defences. If possible, corridors would be cleared through the worst of the defences by a combination of jamming aircraft such as the EF-111 and Wild Weasel defence suppression aircraft, usually F-4 Phantoms.

The maximum external loads that attack aircraft can carry rarely bear much resemblance to the actual

loads that would be carried in war, for several reasons. A maximum ordnance load can severely degrade aircraft performance — the Mach 2-capable F-111, for example, becomes firmly subsonic, its ceiling reduces to less than 15,000ft (4,600m) and its acceleration and powers of manoeuvre, so vital in evading a fighter attack, are severely impaired. Again, missions to attack targets at the limits of the operational radius will often be called for; in this event, external fuel tanks will replace ordnance. And to help the aircraft return safely to base for more missions payload is often sacrificed for protection in the form of either countermeasures pods or air defence missiles such as Sidewinder.

As a rough rule of thumb, the ordnance payload will be about half the stated maximum. Finally, the extra weight and drag of external stores reduces the operational radius considerably. Approximately half the fuel carried externally is used in carrying the remainder to the normal operational radius, leaving only half to increase the endurance.

Weapons carried by attack aircraft are far too numerous for detailed examination in these pages, but generally fall into various groups; guns, old-fashioned iron bombs, smart bombs, unguided rockets, guided powered weapons, anti-runway weapons, area denial and area attack weapons, and, particularly in the anti-surface vessel role, stand-off guided weapons with intertial mid-course guidance and active terminal homing. Finally there are anti-radiation missiles for defence suppression.

The gun carried by an attack aircraft is usually of a type developed for air combat, although it may well fire special munitions such as armour-piercing rounds for the air-to-ground task. It will be of between 20mm and 30mm calibre and will be either a revolving-chambered cannon or a multi-barrel Gatling type. Each has its strengths and weaknesses.

The old but very reliable General Electric M61A Vulcan multi-barrel cannon spews out 100 shells a second, although it does take around three tenths of a second to hit full rate, and slower rates of fire can be

Right: Guns mounted in attack aircraft are usally optimised for air combat but loadings more suitable for ground targets can be used. This is one of the better aircraft guns, the Oerlikon KCA, which has a revolving chamber, electric firing, a high muzzle velocity and a heavy 30mm shell for high lethality.

Below: The 20mm M61A Vulcan cannon has the fastest firing rate of all at 6,000 rounds per minute and is seen here in an F-16 ripping a target to pieces. The shell is light and has poor ballistic qualities: good against soft targets, it is ineffective against armour.

selected; as installed in the A-7D Corsair it weighs 683lb (310kg) without ammunition. It fires a 20mm shell weighing 0.1kg at a muzzle velocity of 3,400ft/sec (1,036m/sec). With its six barrels it is rather a bulky weapon, and its high rate of fire means that it needs a lot of ammunition to give an adequate firing time — including ammunition the A-7D installation weighs 935lb (424kg).

The latest revolver cannon developed for use in the West, and selected to arm the Harrier GR.5, is the British Aden 25. Considerably less bulky than the Vulcan, it weighs some 434lb (197kg), including a full magazine of 200 rounds, so that two can be carried for the same weight as one Vulcan. Rate of fire is between 1,650 and 1,850 rounds per second, with a projectile weight of 0.18kg and a muzzle velocity of 3,445ft/sec (1,050m/sec). Firing time is therefore comparable with that of the Vulcan and if two guns are mounted, as they usually will be, weight of fire is slightly greater. To summarise, weight for weight the American multi-barrel cannon, with its higher rate of fire, is rather better at scoring hits, while Aden 25 hits are far more destructive, the shells being able to penetrate 2in (50mm) of armour at ranges of over 3,281ft (1,000m).

Unique in the aircraft gun world is the gigantic GAU-8 around which the A-10 tank buster is built. This seven-barrel 30mm cannon uses depleted uranium shells to deliver high kinetic energy on the target, is 19.88ft (6.06m) long, and with a full tank of 1,350 rounds weighs 4,029lb (1,828kg). The API projectiles, weighing 0.94lb (0.43kg) each, have a muzzle velocity of 3,240ft/sec (988m/sec) and can be fired at a notional rate of 4,200rds/min — notional because this massive weapon takes over half a second to wind up, by which time the recoil forces have begun to set up a vibration that makes accurate aiming impossible. A typical burst consists of between 30 and 40 shells. The GAU-8 is a very reliable and accurate weapon, with a stoppage rate of once every 22,000 rounds, and an accuracy of 80% shots within 5 mils. (A mil is the span of an object one foot in length viewed from a distance of 1,000ft). From the

Above: The only dedicated tank-killing gun in service is the GAU-8 carried by the A-10A. This picture shows the effect of a single firing pass using a one-second burst of about 50 rounds: marker flags indicate 19 direct hits, while the burnt-out carcass of this M47 attests to the destructive power of the weapon's 30mm shells.

Right: Clouds of smoke spew from the muzzle of the GAU-8 as an A-10A makes a firing pass. The gun fires a depleted uranium-cored shell designed to defeat tank armour with its kinetic energy, at least against the thinner sides or rear. The high muzzle velocity coupled with the great weight of the shell gives a short time of flight, minimising bullet drop.

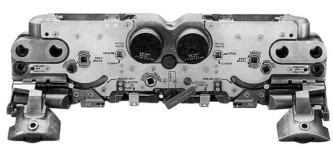

GAU-8's maximum effective range of 6,000ft (1,829m) this would place 80% of the shots within a distance of roughly twice a tank length, giving a good hit probability. Just one hit should be enough.

With the exception of the GAU-8 the aircraft gun is essentially a short-range weapon, effective from little more than 3,000ft (914m) against concentrated soft targets. It also has the disadvantage that the aircraft must be pointed directly at the target, which means it has to run the gauntlet of return fire, and in order to maintain a margin of safety the attack needs to be made at high speed and low level. At a speed of 475kt (880km/h) the aircraft will be covering the ground at over 800ft/sec (244m/sec). Assuming, rather optimistically, that a suitable target can

Above left: BL755 cluster bombs tumble on leaving a CF-18 Hornet. The clean release of stores is always a problem.

Left: Ejector release units like this EDO Model 805 are needed to ensure the safe separation of munitions.

Below: Close support weapons such as these BAT 120 bombs need special adaptors for rapid sequential release.

be seen in a battlefield situation at a distance of around 5,000ft (1,524m), barely three seconds is available for the pilot to line up the gunsight and open fire. This suggests that the only viable targets for guns are those that opportunely appear beneath or very close to the aircraft line of flight, and it is arguable whether an expensive fast jet should be risked in such a way. The gun should be regarded as secondary armament, to be used when all else fails.

The time-honoured air-to-ground weapon is, of course, the bomb, which can be used against a wide variety of targets, both soft and hard. Many types and sizes exist, the usual range for tactical work being between 250lb (113kg) and 2,000lb (907kg), though these are nominal weights and in practice the weapons weigh rather more. The bomb is a very simple and cheap store: the difficulty is landing it in the right place, and the ballistics of bombing would fill a large chapter on their own.

On release, bombs are subject to various aerodynamic forces: the velocity vector of the releasing aeroplane, drag, gravity and the effect of any cross-wind. Release is effected explosively by cartridges in an ejector, to take it clear of the turbulent slipstream, and initially the bomb retains the velocity vector (direction and speed) of the carrier

Matra air-to-ground weapons

Retarded bomb system

Durandal runway-buster

Belouga sub-munition dispenser

Laser-guided bomb system

Rocket pod

aircraft; drag tends to slow it down, while gravity propels it earthward, accelerating at 32.2ft/sec/sec (9.81m/sec/sec/and cross-winds will blow it off course by an amount depending on time of flight and wind strength.

The dropping of bombs is an activity fraught with peril. The target has to be acquired early enough for the aiming data to be fed into the weapons computer, and if target aquisition is visual the aircraft is forced to pull up in order to see it at a sufficient distance, which would be acceptable against a lightly defended target but not against a heavily defended one. In heavily defended areas low-level attacks will be used in conjunction with an accurate nav/attack system and a preplanned attack route. The various methods of bomb delivery, including dive toss, low angle, and laydown, are dealt with in section 3.

Left: This illustration of some Matra weapons covers a spread of target requirements, including general, special-purpose, precision and area attack bombs.

Below: RAF Tornados are seen dropping 1000lb parachute-retarded bombs. Retardation lets the aircraft clear the impact point of the bomb.

Aiming apart, the release of bombs in fast, low-altitude level flight poses problems. Firstly the bombs must have time to arm themselves, so that they will function correctly, but they must not arm themselves too soon, since there is a risk that they will tumble and collide in flight, and they must be prevented from detonating just below the parent aircraft. When the bombs leave the aircraft they are travelling at the same speed, slowing only gradually under the influence of drag. Gravity increases the vertical separation, but only about 16ft (5m) is gained in the first second, so the bomb is effectively flying in close formation with the aircraft. A bomb released at speeds exceeding 600kt (1,100km/h) is capable of flying on body lift alone, which further compounds the problem and often causes restrictions to be placed on release speeds, while the kinetic heating caused by supersonic flight can cause the fuze to malfunction.

The problems of low-level laydown delivery of conventional bombs have been overcome by the use of retarded bombs such as the American Snakeye, which deploy strong air brakes on release to slow them down and allow the aircraft to quickly outrun them. Other methods include the French Matra system, which uses a braking parachute to slow down the bomb and in its super-

retard form permits a release envelope for horizontal flight of 380-600kt (700-1,100km/h) and a minimum altitude of 100ft (30m). An added safety factor with the Matra product is a built-in time delay before detonation of 17 seconds if the speed/-altitude limits are not met.

A conventional bomb works by a combination of impact, blast and fragmentation, and can be used against a wide variety of targets, but its effect is reduced in direct proportion to the distance from the point of impact, and other weapons have been developed for specialised targets or special situations in the form of the runway cratering bomb and the cluster munition.

The most widely used anti-runway weapon is the Matra Durandal. It is small, and weighs just 440lb (200kg), so many can be carried. First used by Israel in the 1967 Middle East War to devastating effect, it is simple and cheap. Released from low level, it is first braked sharply by parachute to an angle which is sufficiently steep to prevent it ricocheting from the hard surface, then boosted by a rocket motor to a speed which enables it to penetrate the concrete runway and burrow below it before detonation. Not only is the runway surface holed, but the concrete is rippled and spalled for a considerable distance around the point of impact, and a cavity is opened up underneath it, compounding the difficulty of repair.

Cluster munitions are of various types according to target, but basically they consist of a container which opens at a preset point after release to scatter a large number of bomblets over a wide area, minimising the inherent inaccuracy of low-level bombing and increasing the possibility of multiple target destruction by a single weapon.

The type of bomblets carried can be selected to give the best results against armour, deployed personnel, transport or supply depots. The Hunting BL755 is fairly typical: in service with 17 nations, it is 8.04ft (2.45m) long, weighs 582lb (264kg) and contains 147 shaped-charge anti-armour bomblets which are ejected at differing velocities to give an even ground pattern. Detonation is on impact, and casing fragmentation is used to give a secondary effect against soft

Above: A line of explosions marks the impacts of a stick of Thomson Brandt BAP 100 runway-busting bombs.

Below and right: The sequence shows the rocket-assisted anti-runway Durandal in action, from release to braking, rocket ignition, impact and detonation.

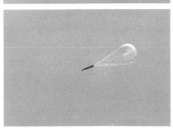

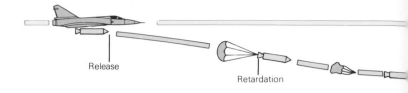

Release

Retardation

targets. CBUs (cluster bomb units) demand both low- and high-speed release limits. Aircraft velocity determines the degree of scatter along the line of flight, while specific release height will give the optimimum scatter pattern; below the optimum altitude lateral scatter will be reduced, and above it there will be a tendency for the bomblets to fall in a pattern which has a hole in the centre.

A potent variation on the CBU theme is the munitions dispenser, examples of which are the British JP233 airfield attack system and the German MW1 area attack weapon. JP233, two of which can be carried by Tornado , is loaded with 30 SG357 cratering munitions, which have much the same properties as Durandal, and 215 HB876 area denial munitions — mines which are retarded to reduce impact with the ground, after which they stand upright on spring steel legs. About 6in (15cm) long and 4in (10cm) in diameter, they are difficult to see among the rubble left by the cratering munitions, and even if seen are dangerous to approach. With shaped-charge warheads, they can destroy armoured bulldozers trying to clear them, the explosion being actuated by sensors, or they can explode on random timing, with considerable fragmentation effect. A single Tornado can deliver 60 cratering munitions and 430 mines.

MW1, which recently entered German service, is designed to deal with concentrations of armour and vehicles and can carry about 4,500 anti-armour bomblets, 2,250 anti-armour bomblets plus 500 anti-tank mines, or 650 fragmentation bomb-

Above: The Belouga dispenser is in widespread use. Operating automatically after release, it carries three different types of submunition—armour-piercing, fragmentation and delayed-action—and a total of 151 bomblets are loaded. Designed for low-level release, the container is retarded by parachute and the submunitions are ejected to cover an area 130ft (40m) wide and either 394ft or 788ft (120m or 240m) long. The spread is selected prior to release.

Right: A JP233 airfield attack dispenser scatters a mix of 30 SG357 cratering munitions and 215 HB876 area denial mines from a Tornado.

Right: MBB's VBW dispenser, now under development, combines IR detection and an 18-tube projectile launcher to give Alpha Jets an autonomous stand-off anti-tank capability.

lets for use against soft targets. The submunitions are fired laterally outwards from the aircraft to give a maximum spread of about 1,650ft (500m) over about the same distance, and the spread can be pre-programmed according to load and target.

So far we have examined air-to-ground weapons which, unless toss-bombing techniques are used, more or less commit the attacking aircraft to overfly its target, inevitably exposing it to defensive fire, but there are other, quite simple weapons, which give a stand-off capability.

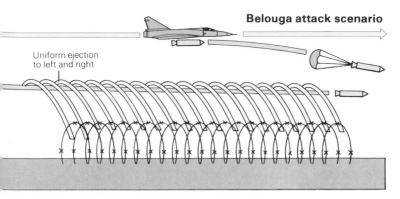

Belouga attack scenario

Uniform ejection
to left and right

VBW MBB

The first and most widely used is the unguided rocket, which is simple and cheap and can be carried in pods and ripple-fired to give a shotgun effect. Typical is the 68mm SNEB rocket, which is spin-stabilised, has a variety of warheads for specific tasks and achieves a speed of roughly 1,165kt (2,160km/h) over and above that of the launching aircraft for an effective range of 3,280-13,000ft (1,000-4,000m), but like all directly sighted weapons it forces the launching aircraft to pull up to acquire its target. Although not particularly accurate, it remains in service, and further developments such as the multidart warhead, which contains a sheaf of steel darts, or flechettes, are under way.

The next step towards a stand-off capability, at the same time improving accuracy, is the laser guided bomb. The theory is simple: a beam of light from a laser can be projected onto a target with great accuracy from long range and the reflections can be detected by the guidance system of the bomb. Designation can be from the launch aircraft, another attack aircraft or the ground, and if released on a trajectory that takes it into the area of reflected laser light the bomb's guidance system will home on the target.

This form of guidance is relatively cheap and very simple, but demands that the bomb be released on the correct trajectory: the bomb has no form of power, and if it lacks the kinetic energy to take it to the target there is no way of stretching its flight. On the other hand, when correctly delivered, it offers a high degree of accuracy, and can be used against pinpoint targets such as bridges and command centres.

More capable air-to-ground weapons are vastly more expensive. Moves are afoot to develop stand-off munition dispensers designed to glide for a considerable distance, using inertial navigation, with infrared linescan target seeking, while powered air-launched cruise missiles will increase stand-off ranges still further, but for the present there is the Rockwell GBU-15 glide bomb, a 2,000lb (907kg) general purpose weapon with either a television or an imaging infra-red sensor in the nose.

Top right: A CF-18 Hornet of the Canadian Armed Forces Aerospace Engineering Test Establishment ripple fires a pod of CRV-7 rockets. The unguided rocket is a cheap area attack weapon, but it lacks accuracy.

Below: An F-111F banks low over rolling country, showing the Pave Tack pod on its belly. The underwing load consists of 1,000lb bombs equipped with Paveway laser-guidance kits.

Delayed lasing

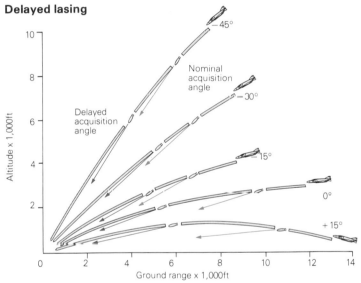

Above: Laser guidance demands a fine balance between early and late target acquisition. Early aquisition with shallow trajectory leads to error, as the bomb noses over too soon, but can be overcome by delayed laser illumination.

Laser-guided bomb loft trajectories

Above: Loft bombing with LGB's gives a stand-off capability, but requires the release point, speed, g force and angle to be accuate and ideally an automatic system should be used. The effects of errors are shown here.

After launch GBU-15 flies a mid-course section using guidance updates provided by data link from the parent aircraft, which will have turned away, while the sensor system will transmit a picture of the area in front of it back to the aircraft where it is shown on a cockpit display; when the target is identified, the Weapons System Officer will either lock on the tracker for automatic terminal homing, manually updating if necessary, or directly guide the weapon onto the target via data link. The next step is a powered version with treble the range of GBU-15 being developed as the AGM-130.

There are many powered guided weapons in the field, but really long range and the ability to penetrate a defended area, identify the target and deliver a worthwhile warload, is probably best achieved by a surface-based cruise missile. A few weapons use laser designation, such as the French AS.30 Laser, while one missile which has been built in enormous numbers, and with several guidance systems, is AGM-65 Maverick, used by a total of 16 nations. A laser-guided Maverick has been ordered by the US Marine Corps for close support work, while the early versions used television imagery and the latest AGM-65D uses imaging infra-red. Quite small — 8.16ft (2.49m) long, 12in (30.5cm) in diameter and, in the D version, weighing 485lb (220kg) — Maverick shows an obvious family resemblance to the Falcon AAM and is supersonic, with a low-level range of about 9nm (16km), though in practice it is likely that weather conditions would limit the range at which the target could be acquired. It can be used against a variety of targets and the TV or IIR sensor presents a picture on the aircraft display: the pilot selects the target, positions crosshairs over it and launches, after which he is free to manoeuvre.

Another family of air-to-ground guided missiles comprises the anti-radiation defence suppression weapons which home on enemy radar emissions. Ground radar is used not only for detection but also to launch and guide surface-to-air missiles and to aim anti-aircraft guns, and if it can be knocked out they

Above: A Mirage F.1 of the French Air Force launches an AS.30 Laser missile. In-flight control is by deflection of the rocket exhaust, giving a twin-motor appearance; mid-course guidance is inertial, with automatic TV tracking and laser illuminated homing.

Centre: AGM-65 Maverick has a launch envelope varying with altitude and speed of the launching aircraft, which determines the distance it can travel. In practice the limit is set by the range at which the pilot can spot the target and lock on the missile.

Right: AGM-88A Harm dives at a radar dish antenna during trials. Designed to home on hostile radar emissions, Harm, or High Speed Anti-Radiation Missile, is used for defence suppression. It can be fired blind towards enemy radars, homing if they switch on subsequently.

Maverick launch zones

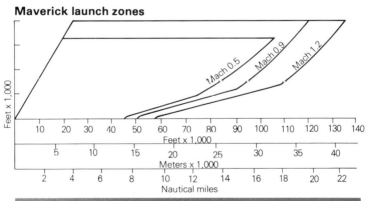

become useless. A typical example is AGM-88A Harm (high-speed anti-radiation missile), which is of the same configuration as AIM-7 Sparrow but rather larger and nearly twice as heavy. Harm can lock on to enemy radar emissions and ride down the beam to the transmitter, and even if it only forces the emitter off the air, thereby saving the aircraft, it has done its job.

Harm can also be launched in a pre-briefed mode, where it is fired in the general direction of enemy radars; if one of them starts working, it will immediately be attacked. The maximum speed of Harm is above Mach 2, and its range is roughly 10nm (18km). The British Alarm performs in much the same sort of way but can also be used in a pre-briefed mode whereby it zooms up to around 40,000ft (12,200m), and then deploys a parachute, enabling it to be suspended above the enemy radars for a protracted period, ready to activate as soon as one of them comes on the air.

Some anti-radar missiles feature inertial navigation which enables them to maintain a steady course for the last known position of a radar which has gone off the air. IIR Maverick can do a good job if teamed with an anti-radiation missile; the standard evasive measure of the ground radar is, shut down if fired at, but IIR Maverick is sufficiently sensitive to be able to home on the heat from the recently used radar.

Currently under development by LTV Aerospace is the Hypervelocity Missile, which carries no warhead but relies on kinetic energy to destroy a target. It is small, weighing less than 49lb (22kg), and cheap, and up to 40 rounds can be carried in a single launcher. Forward-looking infra-red equipment will be used for target acquisition, while guidance is by means of carbon dioxide laser. Initial tests have gone well, proving that it can receive the laser guidance through its exhaust plume, and demonstrations of multiple launch at multiple targets are scheduled for late 1987. The kinetic energy necessary to defeat armour is obtained through the missile's extremely high speed, over 5,000ft/sec (1,524m/sec), or Mach 4.73 at sea level.

Below: A Marineflieger Tornado IDS with a load of four AGM-88A Harm missiles.

Alarm indirect attack mode

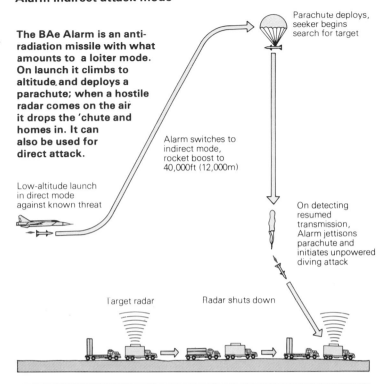

The BAe Alarm is an anti-radiation missile with what amounts to a loiter mode. On launch it climbs to altitude and deploys a parachute; when a hostile radar comes on the air it drops the 'chute and homes in. It can also be used for direct attack.

Parachute deploys, seeker begins search for target

Low-altitude launch in direct mode against known threat

Alarm switches to indirect mode, rocket boost to 40,000ft (12,000m)

On detecting resumed transmission, Alarm jettisons parachute and initiates unpowered diving attack

Target radar   Radar shuts down

The largest guided weapons carried by attack aircraft are, almost inevitably, anti-shipping missiles: ships are large targets and need a destructive warhead to disable or sink them, while range needs to be long by tactical guided missile standards so inertial navigation has to be coupled with some form of terminal homing. There are many types in service, their average weight being around 1,200lb (544kg), with power from either a rocket or a turbojet, speeds in the high subsonic range, operational radii of more than 50nm (93km), and normal approach to the target made at low level.

The target has first to be acquired by radar on the parent aircraft unless the information can be transferred via data link; the target's position is then fed into the navigation system of the missile, which in some cases can be programmed to make a dogleg approach. Such an approach can be very effective, especially if the initial radar contact has been made by an aircraft other than the one launching the missile. After launch the missile flies very low to avoid detection, switching on its active radar only for the last ten miles or so.

Not all missiles use active radar homing; the Norwegian Penguin, which is a smaller missile than most, and is being developed to be carried by the F-16, can use infra-red, which is not a lot of use in fog, but does not betray its presence by an emission. The final approach is usually straight in at low level, but AGM-84 Harpoon uses a pre-programmed pop-up and dive attack, while the British Sea Eagle can be programmed to overfly a ship in order to reach and attack a preselected target.

Air-launched anti-ship missiles first came to prominence in the South Atlantic in 1982 in the form of the French AM.39 Exocet, which has also been used in the Gulf War by Iraq against Iran. Results so far do not seem to be terribly impressive, and the two hits obtained in the South Atlantic appear to have owed a lot to luck. Of course, it is not necessary to physically blow a ship out of the water: the record shows that whatever the weapon used, ships are normally lost to secondary damage caused by uncontrollable fires.

Right: Anti-shipping missiles tend to be large, like this AM.39 Exocet being launched by a Mirage F.1. The target is acquired at long distance by radar, and the coordinates are fed into the missile's inertial navigation system for the mid-course sector; terminal homing is by active radar.

Below: Derived from a surface-to-surface missile, Penguin 3 can be carried by the F-16 as seen here. It can fly a dogleg course if programmed to do so, and uses infra-red terminal homing, which does not betray its presence with an emission and therefore cannot be jammed.

Attack Aircraft

IT is extremely difficult in a work of this nature to produce figures that will enable the reader to make true comparisons between different types of attack aircraft, given the wide variations in external loads carried even by the same type, and in the profiles and natures of the missions they fly.

The average dedicated attack aircraft has six or more hardpoints, each stressed for a different maximum loading and able to carry anything up to a dozen different stores whose weight and drag vary by a considerable amount and both of which affect performance. The mission profile is a factor also. High-altitude flight increases range: in the thin air drag is reduced and the engines burn less fuel, but this can realistically only be used in friendly or neutral areas; in operations over hostile territory low-level flight is the norm, and if afterburner is needed to maintain a high penetration speed, fuel is burned at an alarming rate which reduces range considerably. Range can be increased by carrying fuel externally in drop tanks, but this not only sterilises a hardpoint that might otherwise carry ordnance, thus reducing the weapons load, but tends to operate on a law of diminishing returns, something like half the fuel in the drop tank being used to carry the other half to the point where it materially affects the mission radius.

Many modern attack aircraft have built-in countermeasures, either jamming or expendables, but others are forced to carry them in external pods, which again may sterilise a hardpoint which could otherwise have been gainfully employed, though the Tornado IDS has hardpoints optimised to carry the Skyshadow ECM pod and the BOZ chaff and flares dispenser. It is also becoming more common to equip attack aircraft with air-to-air missiles to give them a credible defence against enemy fighters, and even if this does not utilise a hardpoint designed for air-to-surface weaponry it still adds to weight and drag and in some cases may reduce

the effective weapon load. The performance data given in brochures varies according to both the load carried and the mission profile flown. To quote an extreme example, a clean F-111 can attain Mach 2.2 at high altitude, but with a maximum load of external weapons it becomes firmly subsonic and its ceiling is lower than that of most World War I aircraft.

Range is similarly reduced. The A-7 Corsair is generally quoted as having an operational radius of something in excess of 400nm (740km), but one Vietnam veteran has commented that carrier strikes were generally made from a 'comfortable distance of about 250nm.

The actual performance of any modern military aircraft under any particular set of circumstances is something known only to the manu-

Below: The flat underbody of the Anglo-French Jaguar is well suited to bomb carriage.

facturers and to the services that operate them, and what the manufacturers publish in their brochures, are the best case performance figures. These are of necessity rather bland and give little away. Vmax, for example, is usually stated for the aircraft in clean condition and at high altitude, regardless of the fact that it will rarely if ever be there in a war situation. By the same token, the ability to carry a mere two 500lb (250kg) bombs and deliver them on a target 500nm (925km) away tends to be of academic interest. More to the point is how far it can tote, say, 8,000lb (3,600kg) of bombs and still return to base with an acceptable fuel margin, taking into account that it may have to take evasive action at full throttle on the way in, and fight its way out after the attack.

Flight is a dynamic process and, external loads apart, the capabilities of an attack aircraft are constantly changing throughout the mission. Performance is modified by such factors as speed, altitude and ambient air temperature and pressure, while the weight of the aircraft lessens as fuel is burned. In a typical mission the aircraft would take off heavily laden and cruise-climb to altitude over friendly territory. Once there it would settle to an economical cruising speed, gradually burning off fuel and getting lighter until it approached enemy territory when it would descend to low level to avoid radar detection and jettison empty fuel tanks to reduce drag.

Penetration of enemy territory would be made at high speed and low level, using fuel at a much higher rate. The weapons would then be deposited on the target, lightening the aircraft further and reducing drag, and the egress would be made at high speed and low level in what amounted to a clean condition. Once back in friendly airspace it would cruise-climb back to altitude and return to base at a fuel-burn optimised speed and height. Throughout the mission weight, drag and performance would have changed radically.

The data presented in the form of brochures and press releases is generally accurate for one set of circumstances. We have endeavoured to formulate a simple approach to air-

craft data that will enable the reader to make valid comparisons between different types, without overstepping the constraints of security, by using the non-classified information that is freely available modified with common sense. In the case of Soviet aircraft it has been necessary to hazard an educated guess at some features, although guesswork has been kept to a minimum.

The format adopted for the tabular data and the reasons for its adoption are given where these are not self-evident. All data is given in both Imperial and metric measures, in that order. The aircraft have been listed in chronological order by date of prototype first flight as another aid to making comparisons, and so that development can be followed. Where an aircraft has been developed from an earlier type, this has sometimes been applied, such as the Harrier II/AV-8B/GR.5, but not in others, such as the Su-17/20/22, which was developed from the Su-7 Fitter. The method used is, however, clearly stated in the text.

Dimensions

Length, wingspan and height are given in feet and metres. Wing area is given in square feet and square metres. Also stated is aspect ratio, which affects ride quality at low level.

Weights

Stated in pounds and kilograms, these are often approximate and have sometimes been rounded off. Empty weight is generally the brochure figure where available, while clean take off weight is the weight of the aircraft with full internal fuel and internal guns loaded. Maximum take off weight is the brochure figure to which the aircraft has been cleared, often a paper figure based on the maximum weight that the hardpoints are stressed to carry, and in practice it is impossible to find a combination of weapons that matches each hardpoint exactly. Maximum external load is the sum total of the weights that the hardpoints are cleared to carry, and the number of hardpoints is stated for air-to-ground stores: many attack aircraft have additional points for air-to-air missiles, ECM pods or fuel.

Power

The number and type of the engines are is stated. The thrust generated by one engine is stated in pounds and kilonewtons in terms of static thrust at sea level at both maximum (max) and military (mil) settings. In the dynamic conditions of flight these alter considerably, often giving more than the stated thrust at low altitudes but inevitably reducing as altitude increases, and it should be noted that turbofans (tf) are much more economical at cruise settings than the older turbojets (tj).

Fuel

The quantity carried is stated in pounds and kilograms and for the purpose of standardisation, where volumetric figures only have been available, the weight of fuel has been calculated as being JP-4 at 6.5lb/US gallon. This may lead to marginal inaccuracies in places, but it does mean that like is being compared with like. Internal fuel is always given separately from external fuel. In-flight refuelling capability can be used to extend range to the point where the limiting factor becomes lubricant, oxygen or even pilot fatigue, but it is impractical when even remotely within range of enemy fighters.

Finally, the fuel fraction is the percentage of internal fuel expressed as a proportion of the clean takeoff weight. Figures of 0.27 to 0.30 give aircraft with acceptably long-range performance on internal fuel; below 0.27 they tend to be lacking in operational radius, while above 0.30 they carry a weight penalty not only for the additional fuel, but for the weight of the tanks and the structure needed to carry it. A turbofan aircraft should achieve a better radius of action for given fuel fraction than one powered by a turbojet, although many other factors need to be taken into consideration.

Loadings

Loadings are divided into two areas, thrust and wing. Thrust loadings are expressed as a ratio of static thrust to weight, giving a rough indication of available power comparisons between different types, although it is obvious that it will vary considerably between takeoff with full fuel and full payload and the return to base with minimum fuel remaining and all ordnance expended. Thrust loadings are given for maximum takeoff weight and for clean takeoff weight, the spread between the two giving relative figures with which to compare different types.

Wing loading is stated in lb/sq ft and kg/m² and is calculated over the same two loaded weights, clean takeoff to maximum takeoff. Wing loading is traditionally a measure of instantaneous manoeuvre capability for fighters, but high lift devices have made the relationship more difficult to assess. In any case, a fully laden attack aircraft will not be particularly manoeuvrable, but it will probably have to fly at high speeds at very low altitudes, where gust response is the most important factor.

Gust response is a measure of ride comfort; a low gust response prevents the crew being rattled about too much, a process that can degrade their efficiency considerably. As a general rule a high wing loading coupled with a low aspect ratio gives the lowest gust response and therefore the smoothest ride. The first figure, that for clean takeoff weight, gives a rough idea of manoeuvre capability if the aircraft has to fight its way back to base.

Performance

Maximum speed, or Vmax, is given as Mach number, and is stated both for high altitude, normally 36,000ft (11,000m) and over, and for sea level. These speeds are brochure figures for the clean condition, and as such are irrelevant for all but the homeward journey. As a general rule, all laden attack aircraft are firmly subsonic. The use of afterburner to gain speed simply offers a better target to heat-seeking missiles, consumes fuel more quickly and makes an accurate attack more difficult. When comparing the two figures for Vmax it should be borne in mind that the difference between the speed of sound at high altitude and at sea level is roughly 88kt (163km/hr).

Service or operational ceiling is stated in feet and meters and being for the clean condition also is virtually irrelevant. Initial climb rate, given in ft/min and m/sec, is also a clean condition figure which is attained only at about Mach 0.9 at sea level. Takeoff and landing distances are given in feet and meters. Other performance data is given in the text and suitably qualified.

Below: RAF Tornados with a representative load of bombs, fuel tanks and ECM pods.

BAe Buccaneer

Type: Two-seat twin-engined attack bomber designed specifically for the low level mission. Originally carrier-based but now land-based and used solely for the anti-shipping strike mission.

The Buccaneer was originally conceived for the carrier-based low-level nuclear or conventional strike mission, and despite its antiquated appearance it remains a potent aircraft in that role, lacking only a modern avionics fit. Described by its crews as 'the airborne equivalent of the proverbial brick outhouse' it is a tough, no-nonsense aircraft with low gust response, well suited to the demanding low level flight regime, where it is described as 'running as though on rails'. It cruises quite happily at Mach 0.75 at low level, and its economical Spey turbofans coupled with a high fuel fraction and an internal weapons bay give it a superb payload/range performance.

Designed by Blackburn — now BAe Brough — in response to Naval Requirement NA 39, the Buccaneer was considered to be the answer to the large Soviet Sverdlov class cruisers that were in production in the mid 1950s, delivering conventional or nuclear weapons under the radar.

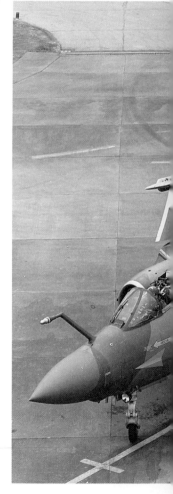

Dimensions	Buccaneer S.2
Length (ft/m)	63.42/19.33
Span (ft/m)	44.00/13.41
Height (ft/m)	16.25/4.95
Wing area (sq ft/m²)	514.7/47.82
Aspect ratio	3.76

Weights	
Empty (lb/kg)	30,000/13,610
Clean takeoff (lb/kg) *	46,000/20,865
Max takeoff (lb/kg)	62,000/28,125
Max external load (lb/kg)	12,000/5,440
Hardpoints	4

Power	2 x Spey Mk 101 tf
Max (lb st/kN)	N/A
Mil (lb st/kN)	11,200/49.8

Fuel	
Internal (lb/kg)	15,612/7,080
External (lb/kg)	3,900/1,770
Fraction	0.34

Loadings	
Max thrust	0.49 – 0.36
Mil thrust	N/A
Wing clean to (lb/sq ft/kg/m²)	89/436
Wing max to (lb/sq ft/m/kg²)	120/588

Performance	
Vmax hi	M = 0.92
Vmax lo	M = 0.85
Ceiling (ft/m)	40,000/12,200 +
Initial climb (ft/min/m/sec)	7,000/36
Takeoff roll (ft/m)	long
Landing roll (ft/m)	long

First flight	May 1963

The low speed requirements of carrier operation were at odds with the needs of high-speed low level flight, and boundary layer control systems were built in to resolve the problem, both on the wings and the horizontal tail surfaces. The extra lift provided reduced approach speed to 124kt (230km/h) while allowing wing area to be kept small to minimise gust response. Other factors affecting low-level performance were an internal bomb bay able to carry four 1,000lb (454kg) bombs, thus eliminating the drag of external carriage, and a blatant example of area ruling which served to reduce drag in the transonic region.

The original Buccaneer S.1 was rather underpowered, with two Gyron Junior turbojets, but the use of Spey turbofans in the S.2 increased the thrust by nearly 60 per cent. In the only export variant, the S.50 for the South African Air Force, the Speys were supplemented by a Bristol BS605 rocket motor for hot and high takeoffs; this gave a further 8,000lb (36kN) of thrust for a period of 30 seconds and is reported to have facilitated some spectacular takeoffs.

During the design phase the identity of the Buccaneer was concealed behind the acronym of ARNA (A Royal Navy Aircraft), a fact which,

Below: The Buccaneer remains a potent aircraft despite its rather ungainly appearance.

coupled with the maker's name of Blackburn, gave rise to the appellation 'Black Banana' or, as it is more often known, 'banana bomber'. Although it was widely recognised as being the best in its class, and rather superior to the American A-6 (its close contemporary), the Royal Air Force refused to buy it, mainly on the grounds that it was not supersonic. One suspects that had it possessed the performance of the F-105 they would have regarded it more kindly, but with afterburning its unsurpassed range performance would have been lost. What other aircraft could launch from a carrier in the Irish Sea and carry out a low level attack on Gibraltar without mid-air refuelling? The round trip on this exercise, staged in May 1966 from *Victorious* totalled some 2,000nm (3,700km).

Finally, in 1968 the RAF relented and ordered the Buccaneer, although following the cancellation of the TSR.2 and F-111 orders there was little other option. It then became manifest what a superior aircraft the Buccaneer was, examples took part in Red Flag exercises at Nellis AFB, Nevada, in 1977, and for the first

three days they successfully evaded both ground and air defences in accomplishing their missions. The type's ultra-low-level transit height — below 100ft (30m) was not uncommon — made it difficult to acquire either on radar or visually, and one aircraft reportedly returned with a tumbleweed hooked on the pitot tube. The Red Force fighters had difficulty in spotting them at tumbleweed height, although they later developed a technique of looking for dust trails; when found, a Buccaneer was usually the cause. This worked in the Nevada desert, but would not have been effective in Europe.

Even when found, at such low levels, missiles could not be relied upon unless the opposing fighter could 'skyline' the Buccaneer, which was difficult, while a gun attack was positively dangerous. One Aggressor F-5E managed to close from astern, no mean feat in itself, only to hit the slipstream of its target and be turned

Below: Four Buccaneers in a typical combat formation at high speed and low altitude over St George's Channel.

Above: The future role of the Buccaneer will be maritime strike, with Sea Eagle missiles as here as its main weapon.

on its back. At that low altitude, it made the pilot's eyes water. Even afterburning fighters had difficulty in catching the elusive Buccaneer — after a short spell at max power they had to break off and seek a tanker.

The adoption of the Buccaneer by the RAF was a purely interim measure until Tornado entered service, but the latest proposals are intended to keep it in service until 1995 or beyond. Originally some 60 S.2s were to be provided with a new analogue AFCS (automatic flight control system) to provide auto-stabilisation in all three flight axes, plus heading, Mach and altitude holds. Improved ECM equipment and expendables dispensers, INS, passive electronic warfare equipment, digital data link and an updated Blue Parrot radar were also proposed, with the first fully modified aircraft scheduled to fly at the end of 1986, but the quantity was later cut back to 42 aircraft with the upgrade aimed almost entirely towards carrying Sea Eagle in the maritime strike role, the Buccaneer having been

replaced in the overland mission by Tornado. The modifications involve a new INS, an updated radar, the Sky Guardian passive warning system, the ALE-40 chaff and flare dispenser and a new radio system, and the first modified aircraft will enter service in the course of 1987.

The Buccaneer is an old design and can hardly be described as pretty, but it is very popular with its crews and its precise handling at high speed and low level is a by-word. The boundary layer control system calls for a high intermediate power setting during the approach to provide sufficient bleed air, which in turn demands very powerful and efficient air brakes. In combat the air brakes can be used to force an attacker to overshoot, being capable of causing a speed loss of up to 20kt/sec.

For self defence Sidewinders are normally carried; there is no gun. The type's only real drawback is that in landing configuration with BLC selected, air brakes out, flaps down and ailerons drooped, the precise handling is lost and it becomes heavy and ungainly with marginal stability, and in need of very careful handling.

Users
South Africa, UK

43

McDonnell Douglas F-4 Phantom

Type: Two-seat twin-engined multi-role fighter employed primarily in the attack role but still used by some countries as an interceptor/air superiority fighter; reconnaissance versions have been developed and the USAF has a Wild Weasel defence suppression variant.

The Phantom is widely regarded as the world's most versatile fighter. It was first introduced into service as a fleet air defence interceptor with the US Navy, and its performance quick-ly aroused USAF interest; after intensive evaluation, during which the Phantom showed itself better than any fast jet then in Air Force service, a modified land-based variant, the F-4C, was ordered. The Phantom was fast, with a good rate of climb, and had good endurance and range by the standards of the early 1960s, though it has since come to be considered fuel-limited. It was also very strong, and soon showed that it could carry a quite extraordinary weight of external stores. With hind-

Dimensions	F-4E	F-4F	F-4G
Length (ft/m)	63.00/19.20	63.00/19.20	63.00/19.20
Span (ft/m)	38.33/11.68	38.33/11.68	38.33/11.68
Height (ft/m)	16.25/4.95	16.25/4.95	16.25/4.95
Wing area (sq ft/m²)	530/49.25	530/49.25	530/49.25
Aspect ratio	2.77	2.77	2.77
Weights			
Empty (lb/kg)	29,535/13,400	28,400/12,880	31,000/14,060
Clean takeoff (lb/kg)	43,150/19,570	41,400/18,780	44,600/20,230
Max takeoff (lb/kg)	61,795/28,030	60,630/27,500	61,795/28,030
Max external load (lb/kg)	18,645/8,460	19,230/8,720	17,200/7,800
Hardpoints	9	7	9
Power	2 x J79-17 tj	2 x J79-17 tj	2xJ79-17 tj
Max (lb st/kN)	17,900/79.5	17,900/79.5	17,900/79.5
Mil (lb st/kN)	11,870/52.8	11,870/52.8	11,870/52.8
Fuel			
Internal (lb/kg)	13,020/5,900	12,400/5,625	13,020/5,625
External (lb/kg)	8,710/3,950	8,710/3,950	8,710/3,950
Fraction	0.30	0.30	0.30
Loadings			
Max thrust	0.83−0.58	0.86−0.59	0.80−0.58
Mil thrust	0.55−0.38	0.57−0.39	0.53−0.38
Wing clean to (lb/sq ft/kg/m²)	81/398	78/381	84/411
Wing max to (lb/sq ft/kg/m²)	117/569	114/559	117/569
Performance			
Vmax hi	M = 2 +	M = 2 +	M = 2 +
Vmax lo	M = 1.19	M = 1.19	M = 1.19
Ceiling (ft/m)	55,000/16,750	55,000/16,750	55,000/16,750
Initial climb (ft/min/m/sec)	28,000/142	28,000/142	28,000/142
Takeoff roll (ft/m)	3,300/1,000	3,300/1,000	3,300/1,000
Landing roll (ft/m)	3,100/950	3,100/950	3,100/950
First flight	Aug 1965	May 1973	Dec 1975

sight this is hardly surprising, as its origins lay in an 'all can do' proposal for the US Navy, one of the main thrusts of which was a single-seat attack bomber.

The introduction of the Phantom to combat came in Vietnam, at first in the air defence, combat air patrol and fighter escort roles, where its lack of an internal gun and poor turning performance often placed it at a disadvantage. Nevertheless, the Phantom gave considerably better than it got in air combat.

Before long the Phantom started taking an active part in air-to-ground operations. The need had been recognised early, and the prototype F-4D, basically an F-4C airframe and engines with upgraded avionics intended to give a better strike capability, flew for the first time in December 1965. The avionics included the Westinghouse APQ-109 partial solid state radar, which gave air-to-ground slant ranging for the first time and featured movable cursors on the display. A lead computing optical sight, a new weapons release computer and a new INS were also fitted, and the lack of an integral gun was rectified by hanging an SUU-23

20mm cannon pod on the centreline. A typical air-to-ground load for this model was 18 750lb (340kg) or 11 1,000lb (450kg) bombs, a considerable weight though less than the rated maximum of the aircraft. For strafing Viet Cong positions — it must be remembered that many air strikes took place in the south of the country; despite the north/south orientation of the conflict, it was not unusual for air units to strike further south than they were based— three 20mm gun pods could be carried. A few F-4Ds remained in service in 1986.

The Phantom was originally regarded as easy to fly, but by modern standards it is not; it takes a lot of practice to become really proficient. Operating in the unforgiving environment of Vietnam, at very high weights, it needed careful handling, and once it departed controlled flight was difficult to recover, needing at least 10,000ft (3,000m) of altitude. Carrying a heavy load of ordnance at high speed and fairly low level, it was

Below: A US Navy F-4B of VF-21 dive-bombs a target in the Vietnamese jungle.

all too easy for the pilot to stray past the limits of control, especially if he was forced to take action to evade ground fire or SAMs, and a lot of Phantoms were lost in this way. Like many swept-wing jets, the Phantom suffers from dihedral effect, a combination of roll with yaw, especially at high AoA, and needs generous amounts of rudder to compensate.

The most numerous of all Phantom variants is the F-4E, the prototype of which first flew in August 1965. It was originally intended to be an F-4D with greater radar capability, but the end product was radically different from what had gone before.

The external gun pod fitted to the C and D was far from satisfactory; in manoeuvring flight the mounts distorted during firing, degrading accuracy, so the F-4E was given an M61 Vulcan cannon with 640 rounds of ammunition housed under the nose.

At the same time, the unhealthy flying characteristics described earlier sparked off a programme called Agile Eagle, intended to give improved handling. The result was that the earlier BLC was omitted, manoeuvre slats were fitted to the wing leading edges and slatted leading edges were fitted to the stabilisers, as had already been done on the US Navy F-4J. More powerful

Above: A GBU-15 glide bomb can just be seen beneath the wing of this F-4E Phantom.

Below: This F-4E has a Pave Tack electro-optical targeting pod on the centreline.

J79-GE-17 engines were fitted, a seventh fuel cell was installed at the rear of the fuselage, and to save weight the hydraulic wing folding was omitted. An improved radar completed the package, which can be regarded as the definitive Phantom, with handling qualities far superior to those of earlier models and later models with unslatted wings.

During the later stages of the Vietnam War Phantoms provided by far the larger part of USAF capability. They carried ordnance, flew barrier patrols, acted as chaff bombers to blind the Vietnamese ground radars and provided fighter escort to attack Phantoms. The US Navy retained its A-6 and A-7 attack aircraft and used the Phantom mainly for fighter escort and barrier patrol, although it was common practice to assign it defence suppression tasks in addition. The first American aces of the war, Lts Cunningham and Driscoll of VF-96, were actually tasked with flak suppression on the mission when they scored their triple kill, armed with two Sparrows and four Sidewinders each for self defence plus six 500lb (227kg) cluster bombs for flak suppression.

The Phantom has also seen considerable action in the Middle East. The first F-4Es entered Israeli service in September 1969 and were used for deep penetration strikes against Egypt in the closing stages of what has become known as the War of Attrition. Equipped with ECM pods, they roved as far as Inshas, Dahsur, and Tel-El-Kebir. They were always regarded as fighter bombers by the Israelis, and were often escorted by Mirages.

The October War of 1973 saw them in the thick of the action against the invading Egyptians and Syrians, and losses were heavy for the first few days, mainly to SAMs and ground fire. At the outset, the Israelis had 128 F-4Es and RF-4Es of which 33 were lost, though replacements received during the three weeks that the war lasted totalled 35. Phantoms accounted for 101 air combat victories during this period, though that is less than a third of total claims and there were almost twice as many Phantoms as Mirages.

Since the Israeli acquisition of the newer and more potent F-15 and F-16 the F-4E has become less important, but considerable numbers remain in service, and they are sufficiently well regarded to have become the subject of an exhaustive and expensive update which includes re-engining.

Below: An Israeli F-4E with a trial installation of two Gabriel anti-ship missiles.

Also in the Middle East, F-4Ds and F4Es have flown for Iran in the ongoing war against Iraq. Many strikes have been carried out, but little of value can be deduced from the meagre information that has been released.

The next Phantom of importance in the attack field was the Luftwaffe F-4F, which first flew in May 1973. Originally ordered as an air superiority fighter, the F-4F was a modified F-4E with simplified avionics, no Sparrow and various weight-saving measures designed to increase the thrust loading and reduce the wing loading, including deletion of the slatted stabiliser, the in-flight refuelling capability and the seventh fuel cell. The F-4F currently equips two fighter-bomber wings in the interdiction and strike roles and is scheduled to receive a comprehensive avionics update to maintain its effectiveness until the turn of the century. Engine modifications to reduce smoke will also be carried out, and a service life extension programme (SLEP) will extend the life of the airframe.

One of the most remarkable members of the Phantom family is the F-4G Wild Weasel. Severe losses to ground fire in Vietnam spurred the introduction of F-105G specialised defence suppression aircraft Later the ubiquitous Phantom took over the role, first flying in December 1975 and reaching initial operational capability in October 1978. The heart of the Wild Weasel is the APR-38 emitter locator system, which detects and classifies hostile radars and SAM systems and presents data to the backseater. The system is very accurate, working by means of triangulation which involves three brief pop-ups from the terrain-hugging altitude at which most of the mission is flown. Once the emitters are precisely located they can be attacked with anti-radiation missiles such as Shrike or Harm (which ride down the emitter's own beam), Maverick missiles or cluster bombs.

The Phantom was chosen for the Wild Weasel role simply because the F-4E happened to be both available and suitable. The internal cannon was deleted to make room for radar detection equipment, but otherwise external variation is limited to the 52

antennas scattered around the aircraft. APR-38 is a comprehensive detection system which as well as locating and classifying hostile emitters provides data for blind bombing, including automatic weapons release. The F-4G also needs to be able to carry out its low-level mission at night and in poor weather; its radar has a terrain avoidance mode, and the navigation equipment has to be very accurate.

The Wild Weasels' main task is to find and destroy hostile radar emitters, principally those associated with SAM systems, and especially mobile systems such as SA-6. Fixed sites can be located and attacked by more conventional means, but the Weasels are a limited asset which should not be risked unnecessarily. Accordingly, they usually hunt in company, a Weasel being paired with a conventional F-4E which it can direct in an attack. If necessary, several F-4Es can accompany one F-4G, and given the right communications and avionics there is no reason why a different type should not be paired with a Weasel.

Barring unforeseen circumstances, it has been estimated that some 2,000 Phantoms will still be flying at the turn of the century. The majority will have undergone some form of improvement — more powerful and economic engines, improved avionics, and fatigue life extensions, as well as new weapons, — and the Phantom will remain a potent force for many years to come.

Users
Greece, Iran, Israel, Japan, South Korea, Spain, Turkey, USA, UK, West Germany

Above: The F-4F Phantom equips two Luftwaffe attack wings; this machine is from JaboG .36.

Below: The WW tail code and extra antennas mark this as an F-4G Wild Weasel Phantom.

Grumman A-6 Intruder

Type: Two-seat all-weather carrier-based attack bomber. Variants are the KA-6 tanker and the EA-6 Prowler four-seat electronic warfare aircraft.

In photographs the Grumman A-6 Intruder has a curiously innocuous look. Its portly but gracefully curving fuselage, the high aspect ratio, almost unswept wing, and the 'Mickey Mouse' divided windshield, combine to give it the appearance of a warplane designed under the auspices of the Disney Studios. But photographs can be deceptive; in real life the Intruder is impressively large

for a carrier-based aircraft and gives an abiding impression of solidity, an impression that is confirmed by its record to date.

The origins of the Intruder lie in a late 1950s US Marine Corps requirement for an aircraft that could hit 'targets of opportunity at night or in marginal weather conditions. The requirement, born of experience in Korea, was intended to produce an aircraft able to locate reinforcements being brought up under conditions that precluded orthodox air attack and accurately place ordnance on them, capable of adverse-weather close support and carrier-compat-

Dimensions	A-6E Intruder
Length (ft/m)	54.75/16.69
Span (ft/m)	53.00/16.15
Height (ft/m)	16.16/4.93
Wing area (sq ft/m²)	529/49.15
Aspect ratio	5.31

Weights	
Empty (lb/kg)	26,600/12,090
Clean takeoff (lb/kg)	43,000/19,500
Max takeoff (lb/kg)	58,600/26,580
Max external load (lb/kg)	18,000/8,165
Hardpoints	6

Power	2 x J52-8B tj
Max (lb st/kN)	N/A
Mil (lb st/kN)	9,300/41.4

Fuel	
Internal (lb/kg)	15,939/7,230
External (lb/kg)	10,050/4,558
Fraction	0.37

Loadings	
Max thrust	0.43 − 0.32
Mil thrust	N/A
Wing clean to (lb/sq ft/kg/m²)	81/397
Wing max to (lb/sq ft/kg/m²)	111/541

Performance	
Vmax hi	M = 0.94
Vmax lo	M = 0.85
Ceiling (ft/m)	42,400/12,900
Initial climb (ft/min/m/sec)	8,600/44
Takeoff roll (ft/m)	4,560/1,390
Landing roll (ft/m)	2,540/774

First flight	1970

ible. Long range and/or extended loiter time were necessary, dictating the high fuel fraction, though at that time such things as fuel fractions had not been thought of.

The mainly low-level attack profile called for was at odds with the low-speed needs of carrier operations; the latter took priority and Grumman settled for a large-span, high-aspect-ratio wing with a high lift coefficient and flaps to almost the entire leading and trailing edges; lateral control was by spoilers. Most unusually, the air brakes are situated on the wingtips and are of a split type, the upper and lower surfaces of the trailing edge opening up and down respectively. Typical approach speed is 120kt (222km/h) and stall speed at normal

landing weight is just under 100kt (185km/h).

The Intruder has been built in several different variants. The original A-6A, equipped with digital integrated attack navigation equipment (DIANE), saw extensive service in Vietnam, flying approximately 35,000 combat missions with both Navy and Marine Corps, the Navy aircraft being carrier-based while the Marine Corps aircraft were based on land at Da Nang and Chu Lai. DIANE was both expensive and unreliable, two factors which contributed to a lack of enough Intruders to perform

Below: A-6A Intruders from USS *Constellation* over the Gulf of Tonkin in July 1968.

all the tasks called for. The expense had resulted in a low acquisition rate, and Intruder squadrons represented only one in five of the fixed-wing squadrons making up the complement of a carrier, while squadron establishment was at first fixed at only nine aircraft (later increased to a dozen).

The unreliability of DIANE was caused by the primitive level of technology, and at one point aircraft serviceability was down to a mere 35 per cent. On the other hand, when it worked it worked very well. The Marine squadrons had an advantage in that they could equip forward air controllers with radar beacons, which gave a precise point on the ground from which the Intruders could offset their attacks. Navy Intruders were used from 1965 until the end of the war in 1973, often on deep penetration strikes and frequently in the monsoon season when no other type could operate. A total of 65 Intruders were lost to enemy action over Vietnam, 47 of them Navy aircraft, but only two were lost to MiGs, both in 1967.

The need to counter the Vietnamese SAM systems gave rise to a defence suppression variant, the A-6B, 19 of which were converted from A-6As to carry the AGM-78 Standard anti-radiation missile. Night attacks against small moving targets such as trucks called for improved detection capability, and a dozen A-6As were fitted with FLIR and LLTV equipment in small turrets

Above: The KA-6D tanker version retains an offensive capability as these VA-52 aircraft show.

under a programme called TRIM (Trails, Roads, Interdiction, Multi-sensor), being redesignated A-6C, and a specialised tanker variant, the KA-6D, was developed.

In the meantime an electronic warfare aircraft developed from the Intruder, the EA-6A Prowler, was an interim design introduced to service in 1967, and only 19 were produced before it was replaced by the greatly modified and far more capable EA-6B, which had two extra seats to accommodate Electronic Warfare Officers and was packed with electronic de.ection and jamming gear. The empty weight of the EA-6B was some 5,500lb (2,500kg) more than that of the basic aircraft, and loaded and landing weights increased in proportion. The airframe had to be strengthened to cope, the wing design was modified to increase lift, the undercarriage was strengthened and more powerful J52-PW-408 engines were fitted.

The current variant of the Intruder is the A-6E. The first deliveries began in 1971, and the changes were mainly to the avionics with the dual goal of improving both capability and serviceability. The two radars of the A-6A were replaced by a single Norden APQ-148 multi-mode radar, with track-while-scan, terrain avoidance and ground mapping modes, while solid-state electronics were introduced, including a new computer and a new nav/attack system. The overall effect was dramatic: system serviceability increased to an average of 85 per cent, navigational accuracy was increased by one third, and bombing CEP (Circular Error Probability) was nearly halved in the radar aiming mode and more than halved in the visual mode. In addition to the new production A-6Es a total of 240 A-6As were converted to the new standard.

In 1976 TRAM (Target Recognition Attack Multisensor) was introduced. The TRAM system contained an imaging infra-red (IIR) sensor, a laser ranger/designator and a laser spot tracker housed in a small turret beneath the radome. The IIR sensor gives a television-quality picture of the target which is displayed on a screen mounted directly above the radar display; capable of considerable magnification, it is an invaluable aid to target recognition and identification at night. Other additions to the A-6E TRAM were an automatic carrier landing system with approach power compensator for fully automatic blind landings at night, a Carrier Aircraft Inertial Navigation System (CAINS), TACAN and improved IFF and communications.

A-6Es have seen limited combat service with KA-6Ds and EA-6Bs they took part in raids over Lebanon in 1983, one being lost in the process, and all three types played a part in the American strike against Libya on April 15, 1986, when Intruders from *Coral Sea* and *America* attacked Benina Air Base and Benghazi Barracks respectively.

The Intruder is pleasant to fly and handles well, though a lower aspect ratio would no doubt make life easier at low level. No credible replacement has yet appeared, and a new variant, the A-6F, is scheduled to enter service in 1990. Again, the main upgrades will involve the avionics, and are expected to double the range of the present system for recognition, acquisition, and tracking: inverse synthetic aperture radar processing will give sharper resolution as well as longer range, while Doppler beam-sharpening will improve ground mapping and enable smaller areas to be examined more closely. The radar will include air-to-air modes, and the carriage of both Amraam and Sidewinder for self defence is proposed. It has also been suggested that Amraam will allow the Intruder to act as an armed airborne picket, though why this might be necessary when Hawkeye and Tomcat are available is puzzling. The cockpit will be changed out of all recognition, with five multi-function displays instead of the previous screens and dials, and the pilot will be provided with a HUD for the first time. The A-6F will be heavier than the A-6E, but will be powered by two F404-GE-400D turbofans, giving 10,700lb (47.5kN) of thrust each without afterburning, which are much more fuel efficient.

User
USA

Below: The A-6E carries a TRAM multi-sensor system whose undernose turrets can be seen.

General Dynamics F-111

Type: Two-seat twin-engined all-weather long-range interdiction and strike bomber. Other variants are the EF-111 electronic warfare aircraft (F-111A rebuilt by Grumman) and the RF-111C operated by the Royal Australian Air Force.

Numerical data for the F-111 tend to consist of extremes. For an attack aircraft it is huge: its empty weight is greater than the maximum takeoff weight of many other types and its maximum takeoff weight, ranging up to 54 tonnes, is enormous. The fuel fraction is on the generous side and

the external fuel that can be carried exceeds the total fuel load of most other attack aircraft, but the weapons load that can be carried is modest in relation to the scale of the aircraft. The internal bomb bay would only be used to house nuclear weapons; otherwise it carries a Pave Tack pod, a 20mm cannon or extra fuel according to type. By modern standards the thrust loading is poor, while wing loadings are exceptionally high. One of the few tactical aircraft to feature side-by-side seating — the Intruder is another — an arrangement which improves crew co-

Dimensions	F-111AE	F-111F	FB-111A
Length (ft/m)	75.54/23.02	75.54/23.02	75.58/23.04
Span (ft/m)	63.00/19.20 max	63.00/19.20 max	70.00/21.33 max
Height (ft/m)	17.04/5.19	17.04/5.19	17.04/5.19
Wing area (sq ft/m²)	525/48.79 max	525/48.79 max	550/51.11 max
Aspect ratio	7.56 − 1.95	7.56 − 1.95	8.91 − 2.10
Weights			
Empty (lb/kg)	46,172/20,940	47,481/21,540	47,980/21,760
Clean takeoff (lb/kg)	79,366/36,000	80,640/36,580	84,957/38,535
Max takeoff (lb/kg)	91,300/41,400	100,000/45,360	119,243/54,090
Max external load (lb/kg)	19,800/8,980	19,800/8,980	19,800/8,980
Hardpoints	4	4	4
Power	2 x TF30-3 tf	2 x TF30-100 tf	2 x TF30-7 tf
Max (lb st/kN)	18,500/82.2	25,100/111.5	20,350/90.4
Mil (lb st/kN)	12,500/55.6	14,500/64.4	12,500/55.6
Fuel			
Internal (lb/kg)	32,715/14,840	32,660/14,815	36,477/16,545
External (lb/kg)	15,613/7,080	15,613/7,080	23,418/10,620
Fraction	0.41	0.41	0.43
Loadings			
Max thrust	0.47 − 0.41	0.62 − 0.50	0.48 − 0.34
Mil thrust	0.31 − 0.27	0.36 − 0.29	0.29 − 0.21
Wing clean to (lb/sq ft/kg/m²)	151/738	154/750	154/750
Wing max to (lb/sq ft/kg/m²)	174/849	190/930	217/1,058
Performance			
Vmax hi	M = 2.2	M = 2.5	M = 2.1
Vmax lo	M = 1.2	M = 1.2	M = 1.2
Ceiling (ft/m)	51,000/15,550	60,000/18,275	51,000/15,550
Initial climb (ft/min/m/sec)	N/A	N/A	N/A
Takeoff roll (ft/m)	N/A	N/A	N/A
Landing roll (ft/m)	N/A	N/A	N/A
First flight	Dec 1964/ Aug 1969	Aug 1971	30 July 1967

operation, it can haul a worthwhile load of ordnance further than any other tactical aircraft and deliver it with precision.

Like the Phantom, the F-111 started out as an all-can-do proposal. Apart from the low-level interdiction and strike roles that it performs so well, it was originally intended to fill the fighter and fleet defence interceptor roles, but rapidly escalating weight during the development stage put paid to such schemes; the poor thrust-to-weight ratios and enormous wing loadings reduced acceleration and manoeuvrability to unacceptable levels for the fighter mission, while the thought of trying to deck-land the 40-tonne monster even on a giant American supercarrier is horrendous.

Cost soared with the weight, troubles were experienced with the engines and intakes, and eventually the expected production run of more than 1,500 was reduced to penny packets of several subtypes as fixes for the various problems were sought. The entire project has been vilified as a can of worms, much of

which could have been avoided if the F-111 had been built from scratch as an uncompromised long range interdiction/strike aircraft.

The F-111 was the first production aircraft in the world to feature variable-sweep wings. Generously endowed with high lift devices, at a minimum sweep (and maximum aspect ratio) they allowed lower take-off and landing speeds and distances than would have been the case with a conventional swept-wing design, while fully swept in the clean condition they permitted speeds in excess of Mach 2 to be achieved, although the poor thrust/weight ratio meant that the F-111 took an unconscionable time getting there. At a 45° setting the reduced aspect ratio and high wing loading gave a low gust response, resulting in a comfortable ride for the crew at very low level, and an acceptable level of fatigue loading for the airframe. Like many other features of the aircraft, the variable-sweep wings, which were manually

Below: Twelve Mk 82 practice bombs fall from an F-111D.

controlled, gave rise to problems, and some F-111s were lost before the causes could be identified and corrected.

The other field in which the F-111 was a pioneer was that of automatic low-level terrain-following flight. Coupled with an extremely accurate navigation system this allowed penetration of heavily defended areas to be made below the radar, at night and in adverse weather, and to take maximum advantage of the capability, a first-pass blind bombing system was added.

The introduction of the F-111 to combat came in March 1968, when a detachment of six aircraft deployed to Takhli, in Thailand, in the Combat Lancer evaluation. The result was startling; no fewer than three aircraft were lost in the first month, all ap-

Above: An F-111 demonstrates terrain-following flight by skiing up a mountainside.

Below: An F-111A takes off from its Thai base for one of the early Combat Lancer missions.

parently to structural failure, and while the total of 55 missions flown clearly demonstrated that the Vietnamese ground radars were unable to detect the low-flying intruders, the defences were ineffectual and the bombs were landing on target, half the detachment was missing. The survivors were withdrawn and the faults rectified.

Four years later, in September 1972, two squadrons of F-111s returned to Takhli, and after a daunting start, when one aircraft failed to return from the first mission, they eventually chalked up some 3,500 sorties with minimal losses. While some strikes were flown against the Khmer Rouge in Cambodia and others against the Ho Chi Minh trail, the great majority were deep penetrations into North Vietnam against what had become the densest and

most experienced air defence in the world. Gone were the days when radar detection could be avoided — the warning lights were illuminated for minutes at a time on some sorties — but the F-111s got through, usually at night, and deposited their munitions, for the most part accurately on target.

Examination of the basic data gives little idea of the true capabilities of the F-111 or its operational employment. Different types have different modes of attack and carry different weapons, though low-level terrain-following flight features in all mission profiles.

The F-111A and E operate mainly with iron bombs, cluster bombs or Durandal runway-busting bombs. The avionic systems fitted to these models are basically of original 1960s vintage, and while the munitions car-

Above: An F-111A of the 366th TTW is seen over Nellis with a load of 24 Mk 82 slicks.

Below: An F-111F with Pave Tack on the centreline and four Paveway laser-guided bombs.

ried can be dropped with a high degree of accuracy, the destruction of small, hardened targets is best left to aircraft carrying precision guided munitions; free-fall weapons lack the accuracy that makes the difference between success and failure.

The F-111F, by contrast, carries the Pave Tack laser designator pod, which can generate an infra-red picture of the target on a cockpit display, enabling the Weapon System Officer to align the laser designator on the target and release Paveway guided bombs with a high degree of accuracy. It was Paveways that were

used against Libya on the night of April 15, 1986, and video pictures of the IR display were broadcast around the world afterwards. This attack, which was remarkable for its accuracy, was a demonstration of the capability of the F-111F, which undertook a round trip of more than 4,000nm (7,400km) with the aid of inflight refuelling. Pave Tack, like other IR systems, cannot penetrate cloud or haze very well, and for operations in such conditions it would probably revert to the same type of loads as the F-111E. The F-111F is equiped with a much more advanced and accurate navigation system and is therefore better fitted to carry out pinpoint strikes on small targets than other models. Pave Tack requires a lot of intensive training if it is to be used

Below: Tripoli, April 14, 1986: Il-76s as seen by Pave Tack on an F-111F three seconds (upper) and one second from target.

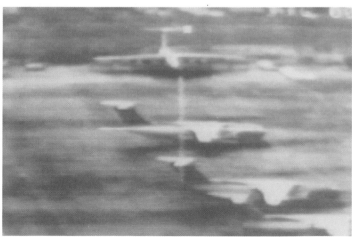

with proficiency, which along with the extra cost of the system is the reason it is not more widely used.

In United States service all F-111 models except one are assigned to Tactical Air Command units. This exception is the FB-111A, which has a bigger wing, longer range, and more lifting power and is assigned to Strategic Air Command as part of America's nuclear deterrent force.

Regardless of the brochure figures, it would be rare for an F-111 to carry a warload of more than 10,000lb (4,500kg), since the added weight and drag would reduce its performance to unacceptable levels and curtail its range severely. Nor would it attempt to use its supersonic capability except in extremis, as afterburning uses fuel at a colossal rate. It can fly at high subsonic speed at low level in military power, and would seek safety by flying fast down among the weeds, preferably at night or in adverse weather. Range is an extremely variable factor, but an F-111 flying a high-low-high mission profile from the UK could reach a target in the west of the Soviet Union quite comfortably. At maximum weights it is firmly subsonic, even using full power, and its ceiling is something below 15,000ft (4,600m).

Users
Australia, USA

Below: One second from weapon impact (upper), then the Pave Tack head swivels to show nine 500lb bombs about to hit.

Nanchang Q-5/A-5 Fantan

Type: Twin-engined single-seat close air support and battlefield interdiction aircraft with limited air superiority capability. The Q prefix denotes the Chinese service aircraft while the A prefix indicates the export models.

Only recently has any firm information on the Nanchang Q-5/A-5 come to light as a result of China exploring the export market. It was known that it had been developed from the F-6 fighter, which in turn was a Chinese copy of the Russian MiG-19: what came as a surprise was the age of the project, the prototype having made its first flight as far back as June 1965, rather than 1972 as had been commonly supposed.

The MiG-19/J-6 Farmer was a highly swept design — about 60° on the leading edge — that was long on thrust and short on fuel. Dating back to 1953, it was the first service aircraft to have a thrust/weight ratio nudging

Dimensions	A-5III	A-5M
Length (ft/m)	50.58/15.42	50.58/15.42
Span (ft/m)	31.83/9.70	31.83/9.70
Height (ft/m)	14.82/4.52	14.82/4.52
Wing area (sq ft/m²)	301/27.95	301/27.95
Aspect ratio	3.37	3.37
Weights		
Empty (lb/kg)	14,316/6,494	14,625/6,634
Clean takeoff (lb/kg)	21,553/9,776	21,863/9,917
Max takeoff (lb/kg)	26,450/12,000	26,450/12,000
Max external load (lb/kg)	4,480/2,032	4,480/2,032
Hardpoints	8	8
Power	2 x WP-6 tj	2 x WP-6A tj
Max (lb st/kN)	7,165/31.8	8,267/36.7
Mil (lb st/kN)	5,730/25.5	6,614/29.4
Fuel		
Internal (lb/kg)	6,353/2,882	6,353/2,882
External (lb/kg)	3,984/1,807	3,984/1,807
Fraction	0.29	0.29
Loadings		
Max thrust	0.66 – 0.54	0.76 – 0.63
Mil thrust	0.53 – 0.43	0.61 – 0.50
Wing clean to (lb/sq ft/kg/m²)	72/350	73/355
Wing max to (lb/sq ft/kg/m²)	88/429	88/429
Performance		
Vmax hi	M = 1.12	M = 1.20
Vmax lo	M = 0.99	M = 1.00
Ceiling (ft/m)	52,000/15,850	52,500/16,000
Initial climb (ft/min/m/sec)	N/A	N/A
Takeoff roll (ft/m)	4,100/1,250	3,940/1,200
Landing roll (ft/m)	3,480/1,060	3,480/1,060
First flight	5 June 1965 (prototype)	

unity at combat weight, and as such it was almost two decades ahead of its time, but the Soviet Union quickly junked it in favour of the Mach 2-capable MiG-21, and it slid into obscurity, only later emerging as a very capable close combat fighter in Vietnamese and Pakistani service.

The Q-5/A-5 Fantan has already appeared in three versions and is about to emerge in a fourth. Ordnance could be hung externally on the J-6 Farmer, but the resulting drag would restrict the operational radius of a fighter already handicapped by the very low fuel fraction of 0.22. The Chinese answer was to redesign it to accommodate an internal weapons bay, and as the F-6 was already a

tight piece of packaging it was decided to extend the fuselage in length and perform an extensive rhinoplasty operation, deleting the simple pitot intake of the fighter in favour of two side intakes. These considerable structural changes increased the weight, and the wings were enlarged by just less than 8 per cent in span and 12 per cent in area. The flaps were redesigned and the underwing spoilers omitted, while the area of the vertical tail was increased and the 30mm cannon was replaced by 23mm cannon of unknown type.

The avionics could only be described as basic on the first aircraft, the Q-5I, which had a radio, Odd Rods IFF and a basic weapons sight that could be used for all functions from air-to-air gunnery to dropping bombs. Virtually nothing is known about the Q-5II: it has been suggested that this is the version exported to North Korea, but exactly what the differences from the I were is not known. The Q-5III/A-5III is the current production model, and is being purchased in large numbers by Pakistan, which also operates the J-6. This model carries its air-to-ground weapons externally; the internal weapons bay has been converted to hold fuel, which brings the fuel fraction up to a more respectable 0.29; for self defence a Sidewinder or Magic heat-seeking missile can be carried on outboard wing stations to augment the guns.

The proposed next model, the export A-5M being developed in a joint venture by CATIC and Aeritalia, has the more powerful WP 6A engines, a slightly better performance than the 5III and a much improved though still basic nav/attack system which includes radar ranging, INS, HUD and air data and central computers. Attack modes include level bombing, dive bombing and dive toss.

Users
North Korea, Pakistan, China

Above left: The Nanchang Q-5/A-5, a MiG-19 extensively modified into a basic attack aircraft.

Left: The Fantan also carries out the maritime strike role armed with two C-801 missiles.

LTV A-7 Corsair

Type: Single-seat single-engined carrier- and land-based attack fighter. Variants include two-seat trainers.

The A-7 Corsair II was designed as a carrier-based attack aircraft to replace the A-4 Skyhawk, which was considered to have an inadequate payload/range capability and little development potential. Bearing a considerable resemblance to the earlier F-8 Crusader fighter, although commonality was almost nil, the Corsair received the appellation SLUF, or Short Little Ugly Fellow, which probably originated from Crusader pilots

and referred to its truncated Crusader appearance. The name was adopted almost as a term of endearment by those who flew it, however.

Early Corsairs were powered by the Pratt & Whitney TF30 turbofan and carried two 20mm Colt revolver cannon. Survivability was a keynote — the cockpit was armoured and essential flight systems were duplicated and widely spaced — while the avionics were sophisticated for the day, with a multi-mode radar, navigational computers, weapons aiming computers and rolling map display. The Corsair soon proved its worth, being able to carry the same weapon

Dimensions	A-7D	A-7E
Length (ft/m)	46.13/14.06	46.13/14.06
Span (ft/m)	38.73/11.80	38.73/11.80
Height (ft/m)	16.13/4.92	16.13/4.92
Wing area (sq ft/m²)	375/34.85	375/34.85
Aspect ratio	4.0	4.0
Weights		
Empty (lb/kg)	19,781/8,975	18,800/8,530
Clean takeoff (lb/kg)	30,000/13,608	29,000/13,155
Max takeoff (lb/kg)	42,000/19,050	42,000/19,050
Max external load (lb/kg)	15,000/6,800	15,000/6,800
Hardpoints	6	6
Power	1 x TF41-A-1 tf	1 x TF41-A-2 tf
Max (lb st/kN)	N/A	N/A
Mil (lb st/kN)	14,250/63.3	15,000/66.7
Fuel		
Internal (lb/kg)	9,600/4,355	9,600/4,355
External (lb/kg)	7,800/3,540	7,800/3,540
Fraction	0.32	0.33
Loadings		
Max thrust	N/A	N/A
Mil thrust	0.48 − 0.34	0.50 − 0.36
Wing clean to (lb/sq ft/kg/m²)	80/390	77/377
Wing max to (lb/sq ft/kg/m²)	112/547	112/547
Performance		
Vmax hi	N/A	N/A
Vmax lo	M = 0.92	M = 0.92
Ceiling (ft/m)	42,000/12,800	42,000/12,800
Initial climb (ft/min/m/sec)	15,000/76	15,000/76
Takeoff roll (ft/m)	< 4,000/1,200	< 4,000/1,200
Landing roll (ft/m)	N/A	N/A
First flight	Sep 1968	N/A

load twice as far as the Skyhawk, or double the weapon load for the same distance, with greater accuracy at night or in adverse weather. One great advantage was its stability in the attack run, which assisted accurate weapons aiming.

It is noticeable that in the 1960s the US Navy was procuring aircraft that were in many ways superior to those in the USAF inventory. That was the case with the Corsair, and the upgraded A-7D for the Air Force first flew in September 1968; many of the same upgrades were wanted by the Navy, which ordered the A-7E at about the same time. Both models were powered by the licence-built Rolls Royce Spey unaugmented turbofan under the designation TF41-

A-1 for the USAF, and the slightly more powerful -2 for the US Navy.

The two 20mm cannon were replaced by the M61 Vulcan cannon with 1,032 rounds, while both versions were given new and more potent nav/attack systems which differ considerably between the D and E. Basically consisting of the Texas Instruments APQ-126 multi-mode radar, a HUD, an INS coupled with a Doppler navigation kit, a stores management panel and various warning and ECM systems, with terrain avoidance and automated attack being two of the new modes available,

Below: An A-7D of the 23rd TFW from England AFB, Louisiana, drops Snakeye retarded bombs.

the Air Force and Navy versions varied so widely that the A-7E was some 1,000lb (454kg) lighter than the D.

The Corsair flew more than 100,000 sorties in Southeast Asia, both from carriers and from land bases, and losses totalled 58, of which all but four were USN aircraft. This reflects the fact that while the Navy Corsairs were extensively used in strikes against the north, those of the Air Force were mainly employed for close air support missions against the Viet Cong; they were also intensively used against Khmer Rouge forces around Phnom Penh, the capital of Cambodia.

Not one Corsair was lost to MiGs, and few fell to SAMS; most were downed by anti-aircraft fire. In action they were renowned for their close air support capability, and on at least one occasion Corsairs hit enemy troops within 75ft (23m) of friendly forces. One factor in their low loss/sortie ratio was undoubtedly their capability to accurately deliver ordnance from a jinking approach, only straightening to aim at the last moment. Corsairs were also used over Lebanon in December 1983, when one was lost to ground fire.

Upgrading has been a continuous process: some Corsairs have been retrofitted with manoeuvre flaps, while others have received FLIR and

Below: An A-7A is "fed to the cat" on board USS *Constellation* off Vietnam, August 1968.

various other modifications. LANA (Low Altitude Night Attack) comprises the addition of FLIR and automatic terrain-following to a total of 48 Air National Guard A-7s to give them a low-altitude approach from beneath the radar coverage, coupled with target aquisition and attack at hight, while six US Navy two-seaters are to be up-engined with the TF-41 and equipped for electronic warfare, though the precise nature of their functions has not been made clear.

The remaining USAF aircraft, numbering some hundreds, have been turned over to the Air National Guard, while the Navy is fast phasing out its Corsairs in favour of the multi-role Hornet. The Corsair was built tough, with a fatigue life of 8,000 hours, and there are hundreds of examples, both flying and in storage, which have considerable life in their airframes, a fact which has induced LTV to enter the update market.

Two schemes recently introduced by the Vought Aero Products Division of LTV are the International Corsair III and the A-7 Strikefighter. Corsair III, which to date has found no takers, is a proposed modernisation of A-7B airframes, including an increase in length of 3.08ft (0.94m) and the fitting of an augmented F110-GE-100 engine with twice the rated thrust of the original TF30. An extra 1,100lb (505kg) of fuel would be carried, though that would barely compensate for even a brief use of afterburner, and advanced digital weapons system and communica-

64

tions are proposed. On the other hand, there can be no doubt that the extra power would come in handy in some situations, shortening the take off run and giving better climb and turn rates, to say nothing of acceleration, although Vmax would be limited by the thick, high-lift wing to Mach 1.12.

The Strikefighter is similar in concept to the Corsair III, and has been developed as a contender for the USAF Close Air Support/Battlefield Air Interdiction (CAS/BAI) role currently filled by the A-10. Powered by either the F110 or the broadly similar Pratt & Whitney F100-PW-200, it would feature a comprehensive avionics suite and a highly modified wing with LEX and a trailing edge flap augmentor, plus automatic manoeuvre flaps, allowing a 4,000lb (1,814kg) increase in maximum take-off weight, more than halving the takeoff roll and greatly reducing the landing roll. Acceleration time from 400 to 550kt (740 to 740-1,020km/h) would be improved by a massive 475 per cent and serviceability would be increased. Sustained g at sea level and Mach 0.8 would also benefit, increasing from less than three to just over six with two 1,000lb (454kg) bombs aboard. It is proposed that no fewer than 462 Corsairs be upgraded in this way, 336 ANG A-7Ds and Ks and 96 US Navy A-7Es.

Users
Greece, Portugal, USA

Above: Shrike-armed A-7Es are readied for flight on *Saratoga's* deck during a Mediterranean exercise in January 1986.

Below: An artist's impression of the LTV Strikefighter equipped for the close air support and interdiction mission.

Dassault-Breguet Mirage F.1

Type: Single-seat single-engined multi-role fighter and attack aircraft. Variants include reconnaissance and two-seat training types.

The limitations of the delta-winged Mirage III/5/50 series were an unnecessarily high takeoff and landing speed with correspondingly long

Dimensions	Mirage F.1E
Length (ft/m)	50.00/15.24
Span (ft/m)	27.58/8.41
Height (ft/m)	14.75/4.50
Wing area (sq ft/m²)	269/25.00
Aspect ratio	2.83

Weights	
Empty (lb/kg)	16,315/7,400
Clean takeoff (lb/kg)	24,030/10,900
Max takeoff (lb/kg)	35,700/16,200
Max external load (lb/kg)	12,786/5,800
Hardpoints	5

Power	1 x Atar 9K50 tj
Max (lb st/kN)	15,870/70.5
Mil (lb st/kN)	11,060/49.0

Fuel	
Internal (lb/kg)	7,384/3,350
External (lb/kg)	7,540/3,420
Fraction	0.31

Loadings	
Max thrust	0.66 − 0.44
Mil thrust	0.46 − 0.31
Wing clean to (lb/sq ft/kg/m²)	89/436
Wing max to (lb/sq ft/kg/m²)	133/648

Performance	
Vmax hi	M = 2.2
Vmax lo	M = 1.2
Ceiling (ft/m)	65,000/22,000
Initial climb (ft/min/sec)	41,930/213
Takeoff roll (ft/m)	1,968/600
Landing roll (ft/m)	2,198/670

First flight	Dec 1966 (prototype)

ground rolls, high energy loss during hard manoeuvring and, more important in the attack role, a very bumpy ride at high speed and low levels which, if sustained for more a than few minutes, lowered crew efficiency. After a brief flirtation with variable-sweep wings on the Mirage G and G8, which seemed to be very capable aeroplanes, France then decided that its real requirement was for a Mach 3 interceptor with fixed wings. The requirement for the *Avion de Combat Futur* changed radically between 1972 and 1975, and hardware finally emerged as the Mirage F.2, a two-seat all-weather fighter powered by the American TF30 turbofan. Meanwhile, Dassault had built a single-seat version which was scaled down and wrapped around the current French engine, the Atar 9K. This was preferred to the larger F.2, and was ordered in quantity as the F.1, at first to fill the air interception role.

The Mirage F.1 had, in numerical terms, a performance comparable to that of the Mirage 3, but its orthodox layout, coupled with full-span leading edge flaps and double-slotted trailing edge flaps to the wings, reduced runway requirements and improved turning capability, while the wings' small lifting area and consequent high loading, combined with a modest aspect ratio to give a low gust response, resulted in a smooth ride at high speed and low level, which suited the aircraft well to the attack mission.

The Mirage F.1A, which has ceased production, was optimised for the attack mission, with the Système d'Attaque au Sol (SAS) using the AIDA 2 radar in place of the Cyrano IV of the air superiority version giving an extremely good first-pass strike capability at low level. The Mirage F.1B and D are two-seat training versions of the F-1C and E respectively, the F-1C being optimised for the interception and air superiority roles while the F-1E is a true multi-role aircraft, being fitted with a modern nav/attack system which includes extra modes to the Cyrano IV radar to give continuously computed air-to-ground ranging, penetration contour mapping, supplementary radar navigation functions and blind let-down. These are combined with the SAGEM ULISS 47 INS, new digital computers and HUD to give greater attack accuracy and digital panels for stores management and navigation to ease the pilot's workload. Future proposals are increased bomb loads, the introduction of a stand-off mode for retarded and cluster bombs and a rear camera for damage assessment.

In Iraqi service the Mirage F.1 has seen considerable action against Iran, and is known to have carried out many anti-shipping attacks against neutral tankers in the Kharg Island area, usually with rockets. The Iraqi Mirage F.1 force comprises F.1EQs, F.1EQ-200s, which are capable of in-flight refuelling, and F.1EQ5s, which carry Agave radar in the nose and are compatible with the Exocet anti-ship missile, as well as two-seat trainers. Exocet and in-flight refuelling have made the Mirage much more for-midable in the anti-ship role: with a single Exocet, two ECM pods and two drop tanks and following a lo-lo mission profile, the combat radius is 380nm (700km), or 485nm (900km) with in-flight refuelling. Also using the lo-lo profile and two drop tanks, it can reach out to 325nm (600km) with six 550lb (250kg) bombs.

Users
Ecuador, France, Greece, Iraq, Jordan, Kuwait, Libya, Morocco, Qatar, South Africa, Spain

Above: An awesome display of firepower as a Mirage F.1 fires four Sneb pods at once.

Below: Mirage F.1s release BAT 120 retarded bombs over a simulated armoured column.

Sukhoi Su-17/-20/-22 Fitter

Type: Single-seat single-engined variable-geometry attack fighter with some counter-air capability. Variants include two-seat trainers and a pod can be carried for reconnaissance missions.

The Soviets are noted for their ability to wring the last ounce out of proven designs, but the Su-17/20/22 is surely an extreme example even for them. The origins of the series, code named Fitter, go back to the mid-1950s when the Sukhoi design bureau turned out a ground-attack type whose low gust response at high speed and low level gave a smooth ride and was coupled with great strength. The early versions had two main faults: they needed very long runways and their low fuel fraction combined with thirsty afterburning turbojets gave them poor payload/range performance. It was decided to increase payload and/or range by adopting a variable-sweep wing, which would also im-

Dimensions	Su-17 Fitter-H	Su-22 Fitter-J
Length (ft/m)	51.83/15.80	51.83/15.80
Span (ft/m)	45.92/14.00 max	45.92/14.00 max
Height (ft/m)	15.58/4.75	15.58/4.75
Wing area (sq ft/m²)	432/40.10	432/40.10
Aspect ratio	4.89−2.49	4.89−2.49
Weights		
Empty (lb/kg)	22,500/10,206	21,715/9,850
Clean takeoff (lb/kg)	34,170/15,500	33,115/15,020
Max takeoff (lb/kg)	42,330/19,200	43,900/19,900
Max external load (lb/kg)	8,160/3,700	10,785/4,890
Hardpoints	8	8
Power	1 x AL-21F	1 x R-29B
Max (lb st/kN)	24,700/109.8	25,350/112.7
Mil (lb st/kN)	17,200/76.4	17,635/78.4
Fuel		
Internal (lb/kg)	10,765/4,885	10,765/4,885
External (lb/kg)	5,495/2,490	5,495/2,490
Fraction	0.32	0.33
Loadings		
Max thrust	0.72−0.58	0.77−0.58
Mil thrust	0.50−0.41	0.53−0.40
Wing clean to (lb/sq ft/kg/m²)	79/387	77/375
Wing max to (lb/sq ft/kg/m²)	98/479	102/496
Performance		
Vmax hi	M = 2.09	M = 2.09
Vmax lo	M = 1.06	M = 1.06
Ceiling (ft/m)	Not released	Not released
Initial climb (ft/min/m/sec)	44,290/225	44,290/225
Takeoff roll (ft/m)	"Moderate"	"Moderate"
Landing roll (ft/m)	"Moderate"	"Moderate"
First flight	2 Aug 1966 (prototype)	N/A

prove the short field performance. Matters were complicated by the main gear being housed in the wing, and the pivot had to be set much further outboard than would have been the case in an optimised design. This gave rise to a rather clumsy-looking wing glove at the original 63° leading edge sweep, and a short movable section with sweep varying between 30° and 60°; sweep control was manual. The resulting Su-17 Fitter-C had four more hardpoints, bringing the total to eight, each capable of carrying a 1,100lb (500kg) bomb.

Early examples were fitted with a simple nav/attack system, but later models were progressively upgraded with laser ranging, Doppler radar and

even terrain avoidance radar. All versions except the two-seat trainers carried two NR-30 cannon with 70 rounds per gun, although it has been rumoured that a few late models are equipped with 23mm twin-barrelled cannon. The Su-17 Fitter-C was exported to several countries under the designation Su-20.

Fitter-D had a more comprehensive avionics fit and a slightly longer fuselage to accommodate the additional black boxes. Fitter-F came next in the sequence; it was powered by the afterburning Tumansky R-29BS-300 in place of the previous Lyulka AL-21F-3, which necessitated alterations to the rear fuselage. It is not really known why this change took place; there is little advantage in terms of extra power in either maximum or military thrust, and later variants were built with either engine and sometimes even alternated between both powerplants.

Fitter-D began to be exported as the Su-22. The avionics fit in the export model was very basic, and was responsible for a fair reduction in empty weight. Fitter-E and -G were two-seaters; the next attack version was the Fitter-H, with the Lyulka engine, a taller fin and a ventral strake. On this model two additional hardpoints optimised for the carriage of AAMs were incorporated. Fitter-J has the Tumansky engine, and has been exported as the Su-22, while the latest known variant, Fitter-K, is in service with the Polish Air Force.

About 800 Su-17s are in service with the Soviet Air Force. Another 70 or more serve with the Navy, in the anti-shipping role, for which they carry the AS-7 Kerry missile.

Users
Afghanistan, Algeria, Angola, Bulgaria, Czechoslovakia, Egypt, Hungary, Iraq, Libya, North Yemen, Peru, Poland, South Yemen, Syria, USSR, Vietnam

Above left: A two-seat Su-17 trainer; the instructor needs a periscope for forward vision.

Left: The small area of moving wing is apparent in this study of Egyptian Su-20s.

Saab AJ 37 Viggen

Type: Single-seat single-engined attack fighter optimised for deployed basing, with some adverse weather capability. The AJ 37 is one of a family of four types; the other aircraft are fighter, reconnaissance and two-seat trainer versions.

Sweden has a small population relative to the area of the country and

Dimensions	AJ 37 Viggen
Length (ft/m)	53.48/16.30
Span (ft/m)	34.77/10.60
Height (ft/m)	18.00/5.49
Wing area (sq ft/m²)	495/46.00
Aspect ratio	2.44

Weights	
Empty (lb/kg)	23,150/10,500 (est)
Clean takeoff (lb/kg)	34,450/15,625 (est)
Max takeoff (lb/kg)	40,000/18,145 (est)
Max external load (lb/kg)	13,200/6,000
Hardpoints	7

Power	1 x RM.8A tf
Max (lb st/kN)	25,990/115.5
Mil (lb st/kN)	14,750/65.6

Fuel	
Internal (lb/kg)	9,750/4,423
External (lb/kg)	(est)N/A
Fraction	0.28

Loadings	
Max thrust	0.75 − 0.65
Mil thrust	0.43 − 0.37
Wing clean to (lb/sq ft/kg/m²)	70/340
Wing max to (lb/sq ft/kg/m²)	81/394

Performance	
Vmax hi	M = 2 +
Vmax lo	M = 1.1
Ceiling (ft/m)	55,000/16,750
Initial climb (ft/min/m/sec)	40,000/203
Takeoff roll (ft/m)	1,312/400
Landing roll (ft/m)	1,640/500

First flight	8 Feb 1967 (prototype)

the length of border that it must defend, and its political stance is determinedly neutral. Nowhere is Swedish neutrality more marked than in its determination to remain self-sufficient in the field of combat aircraft. Since 1945 this policy has resulted in a remarkable string of home-brew fast jets, of which the Viggen is the most recent to enter service. Also notable is the Swedish tendency to ignore what other countries are producing and predicting, a tendency that has turned out two aircraft which resemble nothing else flying. The 1950s-vintage Draken could only be described as futuristic, while its successor, the mid-1960s Viggen, can equally be described as unique. But then, the missions that Swedish fighters are called on to fly have no exact equivalent in the Western world.

The design of multi-role fighters inevitably involves compromise. Attack, interception, reconnaissance and air superiority all have different priorities, and optimisation for any one will normally degrade others. This was overcome with the Viggen by producing four different variants — five if the two reconnaissance missions are counted — using a common airframe. The other aspect common to all is the ability to deploy away from fixed airfields, which in time of war would be targeted in advance. This dispersed basing would be very difficult to knock out; for the most part it consists of straight stretches of road, widened and strenthened and provided with command, communication and repair facilities.

To make the most of the dispersed bases, and to enable it to operate from damaged airfields, the Viggen was designed with a remarkable short field capability and can take off in 1,312ft (400m) and land back in 1,640ft (500m). The takeoff distance is not startling by modern standards — it can be equalled by many other types — but where they need at least three times the distance to land back, meaning in practice that while they could take off from a damaged field they could not return to it, the Viggen's landing run is barely more than its takeoff roll. Three factors account for this: a hard, no-flare carrier-style landing with a sink rate rather more than most land-based fighters could accommodate; a precision landing approach system which allows a precise touchdown; and reverse thrust from the engines for braking. Taking into account Sweden's icy conditions in winter, reverse thrust is essential, as any attempt at orthodox braking would be sporty to say the least.

The Viggen was the first fast jet to enter service with a canard configuration. This was adopted to overcome the worst faults of the delta wing; the canards are fixed, with a trailing edge control surface, and while similar in appearance to those on Rafale and EAP they are in fact quite different, those on the latter being all-moving surfaces. As used on the Viggen, the canard foreplanes simply give good STOL performance to a delta-wing layout, adding lift and enhancing wing lift, the delta being adopted for its supersonic flight characteristics.

The Viggen was really a little ahead of its time: a few years later, it might have benefited from relaxed stability and fly-by-wire, which could greatly have improved its manoeuvre qualities. A few years after it entered Swedish service it was evaluated for NATO against the Mirage F.1E and the F-16 Fighting Falcon, and while the Swedish aircraft had better short field performance than the F-16, the eventual winner, it lost out badly in other areas. In the attack mission, carrying six 500lb (227kg) bombs, its low-level radius of action was 257nm (476km) compared to the 295nm (547km) of the F-16, reflecting a lower fuel fraction and a more thirsty engine. Manoeuvre capability with the same load was much worse: at Mach 0.7 at sea level the F-16 had a turn radius of about 4,500ft (1,372m) against the Viggen's 11,000ft (3,353m).

The Viggen does not carry an internal gun, but can have a pod containing a 30mm Oerlikon KCA cannon. This has a good rate of fire and a very high muzzle velocity, which in combination with the massive shell — some 50 per cent heavier than that used by either the Aden or the DEFA — adds up to a lot of destructive potential. Other weapons carried include anti-ship guided missiles and Maverick, but as yet there is no provision for laser-guided or other smart bombs.

User
Sweden

Right: The Viggen was designed to cope with both dispersed basing and icy conditions.

Below: The AJ 37 Viggen uses a canard foreplane to give good short field performance.

BAe Harrier

Type: Single-seat single-engined attack and reconnaissance fighter with vertical/short takeoff and vertical landing capability. Variants include a two-seat trainer, the Sea Harrier fleet air defence fighter and the AV-8A adopted by the US Marine Corps, later jointly developed as the GR.5 and AV-8B described separately.

On the face of it, the ability to take off and land vertically confers remarkable benefits on a tactical fighter. It can be freed from a conventional airfield with its known location and its miles of vulnerable concrete runway and taxiways. It can be deployed to the most unlikely places and moved around very quickly to avoid detection. It can be based close to the battle area, where its speed of reaction can be vital. And even if caught on an orthodox airfield by a surprise attack it can, at light weight, hover-taxi to an undamaged area for fuelling and arming ready for a sortie.

There are, however, certain drawbacks. Vertical take off demands a considerable surplus of thrust over weight, something more than 1.2:1, which naturally restricts the weapon load that can be carried. It would be

Dimensions	Harrier GR.3
Length (ft/m)	47.16/14.37
Span (ft/m)	25.25/7.70
Height (ft/m)	11.25/3.43
Wing area (sq ft/m²)	201/18.68
Aspect ratio	3.17

Weights	
Empty (lb/kg)	12,200/5,535
Clean takeoff (lb/kg)	18,000/8,165
Max takeoff (lb/kg)	26,000/11,795
Max external load (lb/kg)	8,000/3,630
Hardpoints	5

Power	1 x Pegasus Mk 103 tf
Max (lb st/kN)	N/A
Mil (lb st/kN)	21,500/95.5

Fuel	
Internal (lb/kg)	5,060/2,295
External (lb/kg)	5,150/2,335
Fraction	0.28

Loadings	
Max thrust	N/A
Mil thrust	1.19−0.83
Wing clean to (lb/sq ft/kg/m²)	90/437
Wing max to (lb/sq ft/kg/m²)	129/631

Performance	
Vmax hi	M = 0.97
Vmax lo	M = 1.30
Ceiling (ft/m)	50,000/15,250
Initial climb (ft/min/m/sec)	50,000/254
Takeoff roll (ft/m)	N/A
Landing roll (ft/m)	N/A

First flight	13 Mar 1961
	(prototype)

possible to build a large, multi-engined VTOL aircraft, but only at the expense of complexity, weight and more fuel to lift the extra weight.

Seen from close to, the Harrier GR.3 is a small and rather oddly shaped machine. In action, that is to its advantage, as its small size makes it difficult to aquire visually and its hunched profile, accentuated by the laser-ranging nose, provides a measure of aspect deception, making it difficult to see exactly which way it is going at a distance.

The size of the Harrier is limited by the thrust of the Pegasus 103 engine, which dictates the total all-up weight at which it can operate. In its early days the Harrier was denigrated as

'being able to carry a box of matches the length of a cricket pitch' — a totally unfair assessment of course, but it remains true that the Harrier is fairly short-ranged and carries what is by some standards a light payload; on the other hand, a payload that amounts to 44 per cent of clean take off weight is hardly inconsiderable. Criticisms of this nature also betray a lack of understanding of the basic Harrier mission.

The Harrier is normally operated in the short takeoff, vertical landing (STOVL) mode, which enables it to lift much more weight than in vertical takeoff mode. Its primary function in an all-out conventional war would be to interdict reinforcements approaching the other side of the forward line of troops (FLOT) before they could deploy into the battlefield proper. This would only involve a short penetration of enemy airspace — up to 15nm (28km) — and, coupled with the relatively short distance that it would be deployed behind the FLOT, would mean that the entire mission would last 30 minutes or less and, with quick turnaround, the same aircraft could be back over the same target within an hour.

Such a rapid rate of response would only be limited by pilot fatigue, fuel and stores availability and aircraft serviceability, and in peacetime exercises up to 12 sorties a day have been flown by the same aircraft. Provided it survived the defences, the Harrier could deposit a greater weight of munitions on a greater variety of targets in a shorter time than any other type flying, a fact appreciated by the US Marine Corps, which has adopted it as the AV-8A. The Marine Corps' need to provide rapid reaction air support to its troops can be met by no other attack aircraft in service except the Harrier-derived AV-8B.

The Harrier is usually depicted as lurking in the woods in penny packets, ready to strike at targets of opportunity as they arise. Each off-airfield base would need to be stocked in advance with fuel, spares and munitions, and while little is said

Left: The unique qualities of the Harrier allow it to operate from woodland clearings.

about logistics, this would undoubtedly pose a problem. The great advantage would be that, unlike any other tactical fighter, the Harrier would be operating from locations unknown to an adversary, and even if one were located by the enemy and successfully attacked only a small proportion of the force would be lost.

While the Harrier would be dispersed in small numbers, it would be less effective to employ it in small formations. Once a suitable target was discovered, the Harriers would be sent against it en masse, using pairs as the basic element but with only two or three minutes between the pairs. For example, should an enemy armoured division be located before it had chance to deploy, a squadron of Harriers would be able to subject it to a virtual non-stop attack. The close-up basing would reduce or even eliminate the need to carry fuel externally, and all hardpoints could be used for ordnance, a circumstance that would rarely apply to most conventional aircraft. Nor would the munitions load need to be reduced to extend the range, as is so often the case with other attack aircraft. All these factors contribute to minimising the effect of the Harrier's fairly small maximum payload, and make it equal to if not better than orthodox attack fighters.

Another option, which is rarely mentioned, is urban basing. Instead of 'Harrier Holes' being located in woods, they can be hidden even more effectively in towns. Car showrooms would be ideal, with the glass fronts removed, as long as they had sufficient headroom: the Harriers would operate from the hardstanding outside, or even an adjoining road. Supermarkets, with their large car parks, would make a good alternative. Urban basing would have two main advantages; detection by reconnaissance aircraft would be more difficult — a recently used Harrier towed under the trees and camouflaged can still be detected by IR sensors, whereas indoors this becomes far more unlikely; and an urban base will have a good road network and hard surfaces to prevent ground erosion problems.

The Harrier GR.3 first saw action in the South Atlantic in 1982. In a total of 126 sorties three aircraft were lost, all to ground fire and two of them while returning for a second pass at an alerted target. Based far to the east on the aircraft carriers *Hermes* and *Invincible* they lost much of their quick-reaction capability and were of necessity used in penny packets; nor was any defence suppression available. The Harrier has little or no night attack capability, and all attacks were carried out by daylight. Much was learned during this conflict, and all aircraft are now equipped with better countermeasures. Paveway laser-guided bombs were used for the first time, but cluster bombs were the main weapon carried.

Users
Spain (AV-8S), UK (GR.3), USA (AV-8A)

Right: The US Marine Corps use the Harrier for close air support, as the AV-8A.

Below: Rapid rearming at forward bases cancels out the Harrier's payload/range limitations.

SEPECAT Jaguar

Type: Single-seat twin-engined all-weather attack and strike aircraft. Variants include a two-seat trainer and a carrier-based version developed but not proceeded with, while the Jaguar is also used by some nations in the air superiority role.

Back in the early 1960s both the Royal Air Force and the Armée de l'Air were considering the adoption of a supersonic trainer. At the same time, something was wanted to replace the

Dimensions	Jaguar S (GR.1)
Length (ft/m)	50.92/15.52
Span (ft/m)	28.50/8.69
Height (ft/m)	16.13/4.91
Wing area (sq ft/m²)	260/24.16
Aspect ratio	3.12

Weights	
Empty (lb/kg)	16,975/7,700
Clean takeoff (lb/kg)	24,778/11,240
Max takeoff (lb/kg)	34,000/15,425
Max external load (lb/kg)	10,500/4,765
Hardpoints	5

Power	2xRB.172 Adour Mk 104 tf
Max (lb st/kN)	8,040/35.7
Mil (lb st/kN)	5,320/23.6

Fuel	
Internal (lb/kg)	7,213/3,270
External (lb/kg)	6,182/2,805
Fraction	0.29

Loadings	
Max thrust	0.65−0.47
Mil thrust	0.43−0.31
Wing clean to (lb/sq ft/kg/m²)	95/465
Wing max to (lb/sq ft/kg/m²)	131/638

Performance	
Vmax hi	M = 1.6
Vmax lo	M = 1.1
Ceiling (ft/m)	46,000/14,000
Initial climb (ft/min/m/sec)	Not released
Takeoff roll (ft/m)	2,890/880
Landing roll (ft/m)	1,400/425

First flight	8 Sep 1968

Hunter and Mystère in the attack role. To cut a long story short, British Aerospace were teamed with Breguet (later to be taken over by Dassault) to form a company best known by its acronym, SEPECAT, to develop an aircraft based on a French design and powered by engines built multi-nationally under Rolls-Royce leadership. The supersonic trainer was soon forgotten and the Anglo-French team settled down with their respective air forces to develop a light attack aircraft.

Jaguar has rather the look of a collaborative project. It appears to be large, but appearances are deceptive, and a false impression is given by the ratio of length to span and the stalky undercarriage, which makes it stand high off the ground. Twin engines add to the illusion, although the Adours are small and low powered.

The design of the main gear was conditioned by the requirement for rough field capability, to give plenty of clearance of the external stores. The reason for the choice of two small engines is hard to discover; there were plenty of large ones of sufficient thrust to have done the job. Two engines certainly provides an extra measure of safety — if one packs up or suffers battle damage, the survivor should ensure a safe return to base — but it does not double the safety factor; statistically it improves the attrition rate by about 15 per cent, while cynics could say that there is twice as much to go wrong.

The wing is small, with a moderate aspect ratio, and is highly loaded, all of which gives a low gust response. Short field performance is provided by leading edge slats and double slotted trailing edge flaps to increase lift, but more power to blast it off the ground quicker would have been an advantage. The early Jaguars were powered by the Adour Mk 102, which had rather less thrust than the Mk 104 shown in the table; French Jaguar As have retained the original engine, while the British GR.1s have received the uprated model. Jaguar International, the export version, started out with the Mk 104 (804), but many of them have received the Mk 811, which developes 5,520lb (24.5kN) in military thrust and 8,400lb (37.3kN) with full augmentation.

All single-seat Jaguars are fitted with two 30mm cannon internally, Adens for the British aircraft and DEFA for the French, and Jaguar International can carry two Sidewinders on overwing pylons — a most unusual feature. Payload/range falls between that of the roughly contemporary Harrier and the more modern Tornado.

The primary difference between the British and French Jaguars lies in the avionics fit. The French settled for simplicity with a twin-gyro inertial navigation platform, Doppler radar, a laser rangefinder, navigation and weapons delivery computers, and a radar warning receiver. The British avionics fit was very advanced for its time, with a digital/inertial navigation and weapons aiming subsystem (NAVWASS), a laser rangefinder and marked target seeker in a chisel nose, one of the first HUDs to enter service, and a three-gyro inertial platform; a Ferranti RWR was also carried. The weapons aiming system displayed a continuously computed impact point (CCIP, or 'death dot') on the HUD, giving a miss distance of only 50ft (15m) and making it one of the most accurate strike aircraft of its day. Adverse weather capability was coupled with long range: 290nm (537km) flown entirely at low level on internal fuel, or more than half as far again using a hi- lo-lo-hi profile.

Jaguar is often referred to as a pilot's aeroplane, and handling is pleasant and vice-free, even with a heavy external load; at low level the ride is smooth, and visibility from the cockpit is good. It is generally accepted that a bit more poke would not come amiss, and other faults are a high cockpit noise level, which tends to be tiring on a long mission as well as interfering with communications, and a high workload. The workload is only to be expected, given that the systems involved were designed before modem digital technology was available.

To date the Jaguar has seen active service only with the Armée de l'Air, although it has performed creditably during Red Flag exercises, a formation of six causing some raised eyebrows when they gunned down two F-15s in one mission. In French service they have been used for strikes against Polisario guerrillas in Mauretania during 1977-78, losing at least three of their number in the process, and more recently they have been employed against rebels backed by Libya in Chad.

Users

Ecuador, France, India, Nigeria, Oman, UK

Right: An RAF Jaguar in Norway with a temporary camouflage scheme for a winter exercise.

Below: An Armée de l'Air Jaguar A carries a total of 11 bombs distributed between wing and centreline stations.

Mikoyan MiG-27 Flogger

Type: Single-seat single-engined development of the MiG-23 multi-role fighter adapted for the tactical strike and close air support roles.

The MiG-23 Flogger counter-air fighter could reasonably be described as the first Soviet attempt to produce a tactical aircraft with a useful payload/range. As a fighter it was uninspired, and in both avionics and performance it fell short of the American Phantom, the aircraft that

Dimensions	MiG-27 Flogger-J
Length (ft/m)	54.00/16.46
Span (ft/m)	46.75/14.25 max
Height (ft/m)	14.33/4.37
Wing area (sq ft/m²)	325/30.20
Aspect ratio	6.27 − 2.27

Weights	
Empty (lb/kg)	24,250/11,000
Clean takeoff (lb/kg)	34,764/15,770
Max takeoff (lb/kg)	44,312/20,100
Max external load (lb/kg)	8,820/4,000
Hardpoints	5

Power	
	1 x R29B tj
Max (lb st/kN)	25,350/112.7
Mil (lb st/kN)	17,635/78.4

Fuel	
Internal (lb/kg)	9,914/4,500
External (lb/kg)	4,100/1,860
Fraction	0.29

Loadings	
Max thrust	0.73 − 0.57
Mil thrust	0.51 − 0.40
Wing clean to (lb/sq ft/kg/m²)	107/522
Wing max to (lb/sq ft/kg/m²)	136/666

Performance	
Vmax hi	M = 1.6
Vmax lo	M = 0.95
Ceiling (ft/m)	46,000/14,000
Initial climb (ft/min/m/sec)	N/A
Takeoff roll (ft/m)	2,950/900
Landing roll (ft/m)	2,950/900

First flight	c.1970

it was intended to match. A single-seater, it offered an abysmal view out of the cockpit, with little rear visibility and the forward view obstructed by heavy front screen and canopy framing, though Soviet pilots have commented that this is something that they have got used to and can live with.

Flogger does, however, have certain virtues: it is strong, easy to produce, which means that it can be built cheaply and in large numbers, and it has variable-sweep wings which can be manually set to angles of 16°, 45° or 72°. Minimum sweep reduces takeoff and landing speeds, with a consequent reduction in the runway distance required, increases range and endurance, and permits heavier loads to be carried; intermediate sweep gives optimum turning performance; and maximum sweep reduces drag for acceleration and in high-speed flight, while at high speed and low level it reduces gust response and gives a smooth ride. Coupled with a decent fuel fraction and radius of action, Flogger's aerodynamic versatility made it an obvious choice for the attack role.

The first attack variant was the MiG-23BN, essentially the fighter version with a redesigned front fuselage, rather shorter than the original, which vaguely resembled that of the Jaguar and was quickly dubbed 'utkonos', or 'duck-nose' by the Soviet pilots, The cockpit was revised to give a better view downwards throughout the front quarter, and was armoured against ground fire, while the new nose, freed from the necessity of carrying a large air-to-air radar, sloped sharply down and contained a laser rangefinder.

The MiG-23BN was no more than an interim measure, and an extensive redesign resulted in the MiG-27. The Mach 2 capability of the original fighter, which had been retained in the MiG-23BN, was acknowledged to have no operational value at the low levels where attack missions are carried out, and was deleted: fixed engine inlets were adopted instead of the variable ones used previously, along with a shorter and simpler engine nozzle. These changes limited top speed to Mach 1.6, which for operational purposes is still unusable, and produced a considerable saving in weight, which in turn allowed the payload to be increased. The gear was beefed up to cope with the higher maximum weights, and larger wheels and tyres added, which needed bulged doors to accommodate them; it has been speculated that this was also to give a measure of rough field performance.

For the first time hardpoints were added to the movable portions of the wings: these carry drop tanks only, and do not swivel: the tanks can be carried for the early part of the mission then jettisoned when anything other than minimum wing sweep is required. The avionics are more comprehensive than those of the MiG-23BN, and include a Doppler radar, a radio altimeter and a terrain-avoidance radar mounted in the nose. A laser ranger/marked target seeker is also carried, along with what is believed to be air-to-ground missile guidance radar.

The final major change has been the adoption of a 23mm six barrel Gatling-type cannon carried internally on the centreline, replacing the GSh-23 carried by the fighter. It has been reported by some sources that this gun is trainable in elevation, us-

ing range inputs from the laser, and ground speed inputs from the Doppler. That would make it very effective in the strafing role, increasing the shot concentration considerably.

The latest variant, known as Flogger-J, has also been reported to carry trainable gun pods on hardpoints and is believed to carry a more comprehensive avionics suite than the earlier Flogger-D. Although an effective mud-mover in daylight, the MiG-27 is believed to have little or no adverse-weather or night attack capability. Comparisions with Jaguar are inevitable, as in cold figures the mission looks about the same, the radius of action with internal fuel, carrying an ordnance load of 4,400lb (2,000kg) and using a lo-lo mission profile, being around 210nm (390km). In practice, the Soviet aircraft is shorter-legged, less accurate, and less versatile.

Users
East Germany, India, USSR

Right: Plain inlets, mudguards, laser ranger and a generally rough finish can all be seen on this view of a MiG-27.

Below: The MiG-23BN hybrid was the first attack version of the Flogger to appear.

Sukhoi Su-24 Fencer

Type: Two-seat twin-engined long-range strike and interdiction aircraft with some reconnaissance capability.

When considering the Su-24 Fencer, comparisons with the General Dynamics F-111 become almost inevitable: the two aircraft have a similar configuration, are designed to carry out a similar mission and are broadly comparable in capability. The Americans have referred to

Dimensions	Su-24 Fencer
Length (ft/m)	65.50/19.96
Span (ft/m)	56.50/17.22 max
Height (ft/m)	18.00/5.49
Wing area (sq ft/m²)	452/42.00
Aspect ratio	7.06 – 2.52

Weights	
Empty (lb/kg)	41,890/19,000
Clean takeoff (lb/kg)	64,000/29,000
Max takeoff (lb/kg)	87,080/39,500
Max external load (lb/kg)	24,250/11,000
Hardpoints	8

Power	2 x AL-21F
Max (lb st/kN)	24,250/107.8
Mil (lb st/kN)	16,975/71.0

Fuel	
Internal (lb/kg)	22,045/10,000
External (lb/kg)	18,750/8,500
Fraction	0.34

Loadings	
Max thrust	0.76 – 0.56
Mil thrust	0.53 – 0.39
Wing clean to (lb/sq ft/kg/m²)	142/691
Wing max to (lb/sq ft/kg/m²)	193/941

Performance	
Vmax hi	M = 2.18
Vmax lo	M = 1.20
Ceiling (ft/m)	57,400/17,500
Initial climb (ft/min/m/sec)	28,000/142
Takeoff roll (ft/m)	Short
Landing roll (ft/m)	Short

First flight	1970

Fencer as a mini-F-111, but when it is considered that apart from internal fuel and fuel fraction the Fencer is only about 10 per cent smaller than the F-111 in both dimensions and weights, that it is believed to have more powerful engines, and that most figures released on Fencer are DoD estimates and may err on the low side, that hardly seems a valid assessment. The Soviet Union, naturally, is saying nothing.

Like many Soviet designs of the period, Fencer is a variable-geometry aircraft, with manual sweep settings of 16°, 45°, and 68°. The movable outboard wing sections are well equipped with high lift devices — leading edge slats occupying the full span and three-section double slotted trailing edge flaps, the first time that such a combination had been used by a production Soviet aircraft — and low-speed roll control and lift dumping are provided by spoilers just forward of the flaps. Differentially moving tailerons provide roll control at high speed, another first on a Soviet aircraft.

The feature most immediately reminiscent of the F-111 is the cockpit, with the two-man crew seated side by side, though the Sukhoi Bureau did not run to an escape capsule, contenting themselves with orthodox ejection seats. The seating arrangement is often assumed to be a straight steal from the F-111, but that is not necessarily the case. There can be advantages in side-by-side seating: crew communication is certainly easier, while common switches and instruments can be situated centrally, thus avoiding duplication. But those are hardly good enough reasons to broaden the fuselage of a fighter, and thus increase its profile drag, and it seems more likely that the very large scanner for the nav/attack, terrain-following and weapons delivery radar set the fuselage width and the abreast seating followed. A further spinoff is that a fair amount of body lift is available in flight, easing the highly loaded wings.

For many years the engines were thought to be turbofans, as were the engines of the MiG-23, but it now seems almost certain that they are turbojets, which rather ruins all range/payload calculations made assuming the more economical fan engines. Combat radius with 4,400lb (2,000kg) of ordnance plus two drop tanks has been estimated at 300nm (555km) in the lo-lo mission and 970nm (1,795km) using a hi-lo-lo-hi profile, which means that from bases in the western Soviet Union they could reach all of West Germany and most of Holland at low level, and the whole of the British Isles, France, Scandinavia, Italy, Greece and Turkey using hi-lo-lo-hi. Of course, theory is by no means the same as practice: an attack on Britain would have to run the gauntlet of NATO air defences in Western Europe, followed by UK air defences, at high level, which would be easier said than done.

Fencer is believed to carry an internal cannon, probably a multi-barrel type of 23mm or 30mm calibre. What is certain is that it has a comprehensive avionics kit which enables it to mount extremely accurate first-pass blind attacks at night in adverse weather conditions. The radar is a pulse-Doppler multi-mode type, the modes almost certainly including ground mapping and navigation, while terrain-following radar is certainly included. A laser-ranger/marked target seeker is carried, and what appears to be multi-sensor target aquisition and weapons delivery system on the lines of the American Pave Tack is fitted internally. The radar scanner is being roughly 50in (127cm) in diameter.

It is certain that Fencer carries a comprehensive countermeasurers package. Each successive photograph obtained seems to show more antennas flush mounted with the skin, and at the last count the number was approaching two dozen. A radar warning receiver is certainly included, while active jamming and expendables are carried internally.

Fencer, now identified in three models, remains in production, and by the end of 1986 had certainly passed the total output of F 111s. It is the most formidable weapon in the Soviet tactical air inventory, and will remain so for many years to come.

User
USSR

Above: Two massive drop tanks increase the range of the Su-24.

Below: From this angle Fencer looks nothing like the F-111.

Mitsubishi F-1

Type: Single-seat twin-engined attack aircraft with limited counter-air capability, developed from a two-seat supersonic trainer derivative of the SEPECAT Jaguar.

The Mitsubishi F-1 stemmed from a requirement for a supersonic trainer intended to provide the transition from subsonic training aircraft to the supersonic F-104 Starfighters and F-4 Phantoms of the Japanese Air Self Defence Force. It was probably also intended to give the Japanese experience in the design of supersonic aircraft, and the design was based on that of the SEPECAT Jaguar, itself once intended as a trainer, to which it bears a distinct family resemblance. Early trials of the trainer having proved successful, it was decided to develop it into a single-seat close support aircraft.

The resulting Mitsubishi F-1 looks rather like a Jaguar that has somehow gone wrong. It is longer, but has a smaller span and less wing area; the fin has been completely redesigned to be shorter and broader; the stalky gear is gone, replaced by a shorter and more orthodox-looking undercarriage; the dorsal spine has become less angular; and while the front end is reminiscent of the Jaguar

Dimensions	Mitsubishi F-1
Length (ft/m)	58.58/17.85
Span (ft/m)	25.85/7.88
Height (ft/m)	14.69/4.48
Wing area (sq ft/m²)	228/21.19
Aspect ratio	2.93

Weights	
Empty (lb/kg)	14,017/6,360
Clean takeoff (lb/kg)	21,080/9,560
Max takeoff (lb/kg)	30,146/13,675
Max external load (lb/kg)	6,000/2,720
Hardpoints	5

Power	2 x TF40-IHI-801A tf
Max (lb st/kN)	7,305/32.5
Mil (lb st/kN)	5,115/22.7

Fuel	
Internal (lb/kg)	6,565/2,980
External (lb/kg)	4,276/1,940
Fraction	0.31

Loadings	
Max thrust	0.69 − 0.48
Mil thrust	0.49 − 0.34
Wing clean to (lb/sq ft/kg/m²)	92/451
Wing max to (lb/sq ft/kg/m²)	132/646

Performance	
Vmax hi	$M = 1.60$
Vmax lo	$M = 0.80$
Ceiling (ft/m)	50,000/15,250
Initial climb (ft/min/m/sec)	19,680/100
Takeoff roll (ft/m)	4,200/1,280
Landing roll (ft/m)	N/A

First flight	1977

T-2 trainer the rear cockpit is faired in to provide an avionics bay. Despite the increase in length, empty weight has been reduced by nearly 3,000lb (1,360kg). The Adour turbofans have been retained, built under licence in Japan by Ishikawajima-Harima, but the 30mm cannon are replaced by the rapid-firing M61A Vulcan six-barrel 20mm gun with 750 rounds, mounted low on the left side.

The Japanese Self Defence Forces are committed to eschew offensive action in any shape or form, which makes the attack role rather invidious, and the F.1 is known to the Japanese as a support fighter, a cosmetic description if ever there was one. In practice it is assigned to the attack of any seaborne invasion

force, carrying the indigenous ASM-1 anti-ship missile, a launch-and-leave weapon; alternative loads are iron bombs or unguided rocket pods. Typical radius of action with two ASM-1s and a single 186.6Imp gal (830lit) drop tank is 295nm (550km) using a hi-lo-lo-hi mission profile, reducing to about 200nm (370km) for the lo-lo mission. Two Sidewinders are normally carried on wingtip rails for self defence.

The cockpit is cramped by American standards but well suited to the Japanese physique, and handling is reported to be 'docile'. If the F.1 has a fault it lies in the avionics, which are fairly basic. The AWG-12 fire control system incorporates a Mitsubishi Electric radar offering surface target attack, terrain avoidance and terrain mapping modes in addition to the more standard air-to-air modes, but the radar has been criticised as lacking in range performance. Other items fitted are an INS, Tacan and the APR-4 RWR.

In April 1986 it was announced that the F-1 is to receive a service life extension from 3,500 to 4,500 hours, equivalent to about another three years flying time; an autopilot is to be installed, launchers for AIM-9L are to be added, and a stronger windshield, designed to stop a medium sized bird at 500kt (926km/h), will be fitted.

User
Japan

Left: The faired-in rear crew position reduces the F-1 pilot's rear view to nil.

Below: The Mitsubishi F-1 is officially a support fighter.

Fairchild A-10 Thunderbolt

Type: Single-seat twin-engined close air support fighter and tank buster.

The A-10A Thunderbolt II, more commonly known as the Warthog, was originally conceived for the close air support mission, which involves delivering ordnance accurately in close proximity to friendly troops. Subsequently the requirement to kill tanks was introduced, the result being a big, slow flying gun which could double as a bomb truck.

Fast jets are not much use for the close air support mission: they fly fast and low to survive the defences, which means that if they do manage to pick up a target of opportunity, unless they are fortunate enough to be heading straight at it they are unlikely to be able to line up their sights on it for long enough to take accurate aim. Close to the FLOT, targets are going to be deployed, which means widely spaced and probably moving; they may also be in close proximity to friendly forces,

Dimensions	A-10A Thunderbolt II
Length (ft/m)	53.33/16.26
Span (ft/m)	57.50/17.53
Height (ft/m)	14.67/4.47
Wing area (sq ft/m²)	506/47.02
Aspect ratio	6.53

Weights	
Empty (lb/kg)	21,519/9,760
Clean takeoff (lb/kg)	34,660/15,720
Max takeoff (lb/kg)	50,000/22,680
Max external load (lb/kg)	16,000/7,258
Hardpoints	11

Power	2 x TF34-GE-100 tf
Max (lb st/kN)	N/A
Mil (lb st/kN)	9,065/40.3

Fuel	
Internal (lb/kg)	10,650/4,830
External (lb/kg)	11,700/5,310
Fraction	0.31

Loadings	
Max thrust	N/A
Mil thrust	0.52−0.36
Wing clean to (lb/sq ft/kg/m²)	68/334
Wing max to (lb/sq ft/kg/m²)	99/482

Performance	
Vmax hi	M = 0.59
Vmax lo	M = 0.68
Ceiling (ft/m)	45,000/13,700
Initial climb (ft/min/m/sec)	6,000/30
Takeoff roll (ft/m)	4,000/1,220
Landing roll (ft/m)	2,000/610

First flight	May 1972 (prototype)

and mistaken identity is to be avoided, yet indentification takes time that the fast jet driver doesn't have. If the target is missed on the first pass, the fast jet has to turn round, which takes time and space, and in conditions of poor visibility in hilly terrain with a low cloud base — typical European weather in fact — that may not be possible. With the exception of the Harrier, fast jets will generally be based far back from the battlefield, and the time lapse between a call for help and their arrival will often be too long; the situation will have changed before they arrive.

The A-10 was designed specifically to deal with such a situation. It flies slowly enough to give the pilot time to see, identify and take aim at small individual targets such as tanks; it is slow enough to take full advantage of terrain masking; and in hilly terrain, under a low cloud base, it can turn tightly enough to remain in the same area without losing visual contact.

Prolonged flying at low level in close proximity to the enemy means that it is going to get shot at a lot, and without the speed that the fast jets rely upon for survival it is going to take hits. On the other hand, in the immediate vicinity of the battle area little in the way of heavy metal is likely to be encountered; the immediate opposition will be small man-portable SAMs and mobile AA guns such as

Below: An A-10A turns tightly just above the treetops.

the ZSU-23/4 Shilka. While a lethal hit from any of them is possible, recent experience of local wars suggests that many aircraft are lost to just two or three hits. The A-10, accordingly has been designed to be survivable. The pilot sits in a bathtub of titanium under a thick canopy; the systems are all duplicated or triplicated; extraordinary measures have been taken to avoid fire in a stricken fuel tank; and the bird can still be brought home with half a wing, half the tail, or even a engine missing.

Short field performance was written into the specification, but in reality the type's field performance is not particularly good, and the A-10 may have to be based well behind the FLOT. The consequence has been a revival of the World War II cab-rank system, with aircraft loitering behind the lines until called upon, and the long endurance required was achieved by combining a decent fuel fraction with engines having a very low specific fuel consumption, actually high-bypass ratio turbofans developed for airliner usage. Another

feature of the A-10 is repairability, which is essential for an aircraft designed to absorb punishment: large components can be replaced quickly, and on-the-spot repairs can be carried out to superficial damage very quickly.

Iron, cluster and laser-guided bombs can be carried, but the A-10's main guided weapon is Maverick, which employs electro-optical guidance and gives some stand-off capability; the latest version of Maverick, with IR guidance, entered service in 1986. The A-10's main weapon, however, is its gigantic 30mm GAU-8/A Avenger seven-barrelled cannon, around which the aircraft was designed. This gun system has an ammunition capacity of 1,174 rounds, and two rates of fire are available, 2,100 or 4,200 rounds per minute. The muzzle velocity of 3,500ft/sec (1,067m/sec) is very high, and aids accuracy; the dispersion of shot is stated to be 80 per cent

Below: Needing only short strips to operate from, the A-10A can be forward based.

within five mil, or 30ft (9.14m) at 6,000ft (1,829) slant range, the effective maximum range of the gun.

The round is specially designed for tank busting and consists of an aluminium shell surrounding a core of depleted uranium, one of the densest substances known to man; when it hits the armour of a tank it releases a great deal of kinetic energy, which not only causes the armour to spall, resulting in great flakes of it flying around the tank interior at high speed, chewing up anything that gets in their way, but also releases enough heat to kill or incapacitate the crew if the spalling has not already done so.

In action A-10s will tend to hunt in pairs, and one of their tasks will be to neutralise enemy AAA; if this can be located it will be suitable target for Maverick, which is a fire-and-forget weapon. The attack profile will be quick climb, lock-on, and launch from a range of about two miles (4km). If conditions are unsuitable for Maverick the gun must be used, from a range of 4,000-6,000ft (1,200 - 1,800m): a rapid turning climb, nose down, line up and a one second burst should be enough. In fact, one second is about the maximum effective firing period, as the colossal recoil makes it impossible to hold the sight on target for longer.

Manoeuvrability is the keynote to survival for the A-10. At 300kt (556km/h) it has an instantaneous turn rate of nearly 25°/sec and a radius of about 1,200ft (366m), while at 150kt (278/h) the corresponding figures are roughly 15°/sec and 900ft (275m). At low level, and using countermeasures, it is a difficult target even for radar-directed guns.

The A-10's lack of night capability may be remedied when Lantirn enters service in the near future. In the meantime the 30mm cannon provides its only air-to-air self defence capability, but moves are afoot to equip it with Sidewinders.

User
USA

Below: Its ungainly appearance is belied as this fully armed A-10A rolls rapidly inverted.

McDonnell Douglas F-15 Eagle

Type: (F-15A and C) single-seat air superiority fighters with ground attack as a secondary role. (F-15B and D) fully combat-capable two-seaters. The C and D are able to carry the Fast (Fuel and Sensor Tactical) pack, which increases fuel capacity with minimal drag and without sterilising the wing pylons. (F-15E) two-seat attack-optimised variant with all-weather capability. Apart from dimensions, figures for the F-15E are all provisional.

The F-15 Eagle was designed for the air superiority and interception missions in a mistaken attempt to match what were thought to be the capabilities of the Soviet MiG-25 Foxbat. It was given everything that a fighter could possibly need: an advanced avionics fit with the latest computer technology; eight missiles with a mix of IR and SARH, for close and BVR combat; a fast-firing 20mm M61 Vulcan cannon; a moderate aspect ratio and low wing loading for manoeuvrability; and a tremendous thrust-to-weight ratio for rapid climb,

Dimensions	F-15C	F-15C Fast	F-15E
Length (ft/m)	63.75/19.43	63.75/19.43	63.75/19.43
Span (ft/m)	42.81/13.05	42.81/13.05	42.81/13.05
Height (ft/m)	18.46/5.63	18.46/5.63	18.46/5.63
Wing area (sq ft/m²)	608/56.50	608/56.50	608/56.50
Aspect ratio	3.01	3.01	3.01

Weights			
Empty (lb/kg)	29,180/13,236	30,300/13,700	32,000/14,515
Clean takeoff (lb/kg)	44,500/20,815	55,270/25,070	56,970/25,842
Max takeoff (lb/kg)	68,000/30,844	68,000/30,844	81,000/36,742
Max external load (lb/kg)	16,000/7,258	12,730/5,774	24,000/10,885
Hardpoints	9	9	N/A

Power	2 x F-100-PW-200 tf	2 x F100-PW-200 tf	2 x F-100-PW-200 tf
Max (lb st/kN)	25,000/111.0	25,000/111.0	25,000/111.0
Mil (lb st/kN)	16,200/72.0	16,200/72.0	16,200/72.0

Fuel			
Internal (lb/kg)	13,455/6,103	23,205/10,526	23,205/10,526
External (lb/kg)	11,700/5,310	11,700/5,310	N/A
Fraction	0.30	0.42	0.41

Loadings			
Max thrust	1.12 − 0.74	0.90 − 0.74	0.88 − 0.62
Mil thrust	0.73 − 0.48	0.59 − 0.48	0.57 − 0.40
Wing clean to (lb/sq ft/kg/m²)	73/357	91/444	94/457
Wing max to (lb/sq ft/kg/m²)	112/546	112/546	133/650

Performance			
Vmax hi	M = 2.5	N/A	N/A
Vmax lo	M = 1.2	N/A	N/A
Ceiling (ft/m)	65,000/19,800	N/A	N/A
Initial climb (ft/min/m/sec)	50,000/254	N/A	N/A
Takeoff roll (ft/m)	900/275	N/A	N/A
Landing roll (ft/m)	N/A	N/A	N/A

First flight	Feb 1979	N/A	Jul 1980 (prototype)

sparkling acceleration and excellent sustained turn. The top speed was originally intended to be Mach 3, but ultimately it settled at Mach 2.5.

The result was an extremely capable air-to-air fighter, but one which was also very large and fantastically expensive. Then in 1976, when Lt Viktor Belenko defected to Japan with his Foxbat, the MiG's secrets were laid bare to the West: a combination of the crude and the ingenious, it was found to be a pure interceptor with virtually no manoeuvre combat capability, which left the Eagle as the world's best air superiority fighter by quite a margin.

When one has a piece of hardware as expensive as the F-15, the temptation inevitably exists to use it for other purposes than those it was designed for. The surplus of specific excess power and the relatively low wing loading promised a great deal of lifting capability, and it was just a question of time before someone hung air-to-ground ordnance on it and declared it to be a multi-role fighter which would have the added advantage of being capable of fighting its way home after delivering its load.

First, though, some problems had to be solved. The engines were rather touchy at first and had to be flown with care if shock stalls were to be avoided. Also, in spite of having an adequate fuel fraction, the Eagle was found to be somewhat short on endurance, at least partly because afterburner light-up was a bit unreliable, with the result that during combat manoeuvres pilots tended to leave the engine in minimum burner rather than reduce to military power and risk being embarrassed when it failed to relight. Drop tanks could always be used to increase endurance, but they also increased drag, and sterilised pylons that could otherwise hold ordnance. The F-15 has a total of nine hardpoints, but four are optimised for Sparrow AAMs and two are lightly rated and suitable only for carrying jamming pods and suchlike: hang two drop tanks under the wings and only the centreline position was left for the warload in the air-to-ground mission.

Below: The two-seat Strike Eagle, forerunner of the F-15E, with a load of CBUs.

The answer was the conformal Fast pack, which increased internal fuel at minimal cost in drag and with no penalty in store stations, as well as housing sensors such as IR seekers, ECM equipment, laser designators and even reconnaissance cameras. Another advantage of Fast packs was that air-to-ground stores could be carried tangentially along the corners, replacing the Sparrows.

It is ironic, if typical, that a fighter with a high thrust loading and low wing loading should have both of these advantages traded for carrying capacity, which has obviously eroded performance. The ultimate development will be the F-15E, a dedicated two-seat attack aircraft in which it is proposed to push the maximum takeoff weight up by very nearly six tons to 81,000lb (36,742kg). It can be assumed that most of the performance that characterised the Eagle will be lost, and it is a fair assumption that the Fast pack will be standardised.

The front cockpit of the F-15E will be more modern than previous models, with a wide angle HUD and moving map display, but the rear cockpit has been missionised and has no flight controls. Four multi-function displays (MFDs) fill the top of the dash, which allow the Weapons Systems Officer to monitor aircraft systems, weapons status and enemy radar and missile defences while using a TSD (Tactical Situation Display) to give the broad picture, probably linked to JTIDS. He can also use SAR (Synthetic Aperture Radar) to locate previously unknown targets. Carriage of a Pave Tack laser designator pod has been considered, but will probably be rejected in favour of Lantirn, currently undergoing trials. Pre-programmed nav/attack systems similar to those equipping Tornado will be used for accurate first pass attacks at night in poor weather.

The F-15E is officially needed to supplement the F-111 in the all-weather deep penetration role, but even with Lantirn it is questionable whether it will be able to carry out a role which seems impossible without terrain-following radar for low-level flight. The large wing area militates against low gust response, and one

Below: A good view of the Fuel And Sensor Tactical (Fast) pack carried by the F-15.

report refers to a penetration altitude of 500ft (150m), which against modern defences is far too high. Trials have also been carried out using three 30mm GEPOD tank-busting cannon, but the rationale of using such an expensive aircraft in the close air support role against relatively low-value targets must be open to question.

As the weights have crept up, so the thrust loading has fallen. That is about to be remedied, and future F-15s are to be up-engined with either a more powerful Pratt & Whitney or General Electric's F110, rated at 29,000lb (129kN) and 29,500lb (131kN) respectively.

The Eagle has seen considerable combat in the Middle East, and has built up an enviable record in the air superiority role, mainly in Israeli service, but Saudi Arabian F-15s have shot down two Iranian Phantoms to date. Less is known of its attack record, and it seems unlikely that the Israelis, with plenty of Phantoms and Kfirs, to say nothing of F-16s, would have risked it for ground attack in Bekaa Valley in 1982. It does seem, however, that the long-distance Israeli strike against the PLO head-quarters on the Tunis waterfront on October 1, 1985, was carried out entirely by F-15s, using both 1,000lb (454kg) iron bombs and laser-guided weapons. Six Eagles, supported by ECM aircraft and tankers, made the 2,600nm (4,800km) round trip with the aid of three in-flight refuellings.

The Eagle is capable of more development yet. The Wild Weasel role is an obvious possibility, using the system carried by the F-4G Phantom backed by anti-radiation missiles, and a STOL demonstrator is planned to be able to operate from a 1,500ft (457m) runway in wet conditions with poor visibility. The latter will feature canard foreplanes on the intake sides, two-dimensional vectoring nozzles and thrust reversers, features which should not only give good STOL performance but also enhance manoeuvrability considerably.

Users
Israel, Japan, Saudi Arabia, USA

Below: The Lantirn navigation pod, seen here under the fuselage of a Fast-equipped F-15C, will equip the F-15E.

Dassault-Breguet/ Dornier Alpha Jet

Type: Two-seat twin-engined armed trainer/light attack aircraft. Variants include the Alpha Jet NGEA (Nouvelle Generation Ecole-Appui), the Alpha Jet Lancier for close suport and anti-shipping operations, and a Dornier-developed advanced trainer with a state of the art cockpit.

The Alpha Jet was designed to meet

Dimensions	Alpha Jet NGEA
Length (ft/m)	43.42/13.23
Span (ft/m)	29.92/9.12
Height (ft/m)	13.75/4.19
Wing area (sq ft/m²)	188/17.50
Aspect ratio	4.75

Weights	
Empty (lb/kg)	7,749/3,515
Clean takeoff (lb/kg)	11,398/5,170
Max takeoff (lb/kg)	17,637/8,000
Max external load (lb/kg)	5,510/2,500
Hardpoints	4

Power	2 x Larzac 04 C20 tf
Max (lb st/kN)	N/A
Mil (lb st/kN)	3,175/14.1

Fuel	
Internal (lb/kg)	3,648/1,655
External (lb/kg)	3,174/1,440
Fraction	0.32

Loadings	
Max thrust	N/A
Mil thrust	0.56−0.36
Wing clean to (lb/sq ft/kg/m²)	60/295
Wing max to (lb/sq ft/kg/m²)	94/457

Performance	
Vmax hi	M = 0.86
Vmax lo	M = 0.85
Ceiling (ft/m)	48,000/14,600
Initial climb (ft/min/m/sec)	11,220/57
Takeoff roll (ft/m)	1,345/410
Landing roll (ft/m)	2,000/610

First flight	April 1982

a joint need by France and Germany for an advanced trainer, while Germany also needed a light attack/reconnaissance aircraft to replace its ageing Fiat G.91s. The dual requirement was not incompatible — almost any aircraft has some weapons carrying capability, and an advanced trainer more than most — and it is economic sense to operate a dual-role trainer/attack aircraft to increase force size in time of war, although the Armée de l'Air were to use the Alpha Jet purely in the training role, while the Luftaffe intention was to use it operationally with the rear seat removed.

The two seats were set in tandem, with the rear one raised to give the back-seat pilot a good view over the front seater's head. Two small Larzac turbofans were adopted to give twin-engine safety, and a short though sturdy undercarriage, capable of grass field operation, was used, a feature which necessitated the use of a shoulder-mounted wing to give sufficient clearance for a wide variety of stores to be carried. Clearance beneath the fuselage was too little for much other than a gun pod, a 30mm DEFA cannon being carried by the French aircraft and a 27mm Mauser by the German, each with 125 rounds.

As might be expected, the avionics fit on the Luftwaffe Alpha Jets was more comprehensive than that on their French counterparts; apart from the usual communications gear, ILS, Tacan and IFF, a simple computerised weapon aiming sight is carried, while the German aircraft features a HUD, radio altimeter, Doppler velocity sensors and navigation equipment and a far more sophisticated weapons release system. It has a pitot probe, and a more pointed nose than the French trainer, while the Alpha Jets dedicated to the attack role have no duplicated controls in the rear cockpit; instead they carry various items of ECM gear on the seat bearers. The nose shape is the main external difference btween the two, but the French aircraft also feature narrow strakes on the sides of the nose to improve spin resistance.

The next development, the Alpha Jet NGEA, is fitted with a much more capable nav/attack system and has a chisel nose with a laser ranger, a more accurate INS, and a HUD, coupled with the features already adopted by the Luftwaffe machine. Five attack modes are possible, including offset bombing.

The most recent proposal is the Alpha Jet Lancier, which appears to be oriented towards the anti-shipping role if the Agave radar (which equips the Super Etendard) and the carriage of the Exocet missile are anything to go by. The variant is to have all-weather capability and employs FLIR imaging displayed on the HUD. Another interesting feature is that it can carry up to three 30mm DEFA gun pods for the anti-helicopter role, no new mission for the Alpha Jet; the Luftwaffe dedicated attack aircraft would be given this task in war.

The shortage of hardpoints makes it unlikely that the Alpha Jet would carry drop tanks in war as it would lose half its load carrying capacity at a stroke, but with a good fuel fraction and economical turbofans its payload/range is perfectly adequate for its task: in the close air support mission, flown with six 550lb (250kg) retarded bombs and the gun pod, it has a loiter time of 35 minutes at a radius of 110nm (200km) using an all-lo profile, while the same sortie with no loiter time could extend to a radius of 205nm (380km) in a lo-lo profile or 305nm (565km) using a hi-lo-lo-hi profile.

The Alpha Jet is not the most inspiring attack aircraft around, but it provides a modicum of capability at a reasonable cost and is well suited to the needs of the emergent nations that make up the bulk of its overseas buyers.

Users
Belgium, Cameroon, Egypt, France, Ivory Coast, Morocco, Nigeria, Qatar, Togoland, West Germany

Above: The Luftwaffe uses the Alpha Jet A, seen here, mainly in the close air support role.

Below: The Alpha Jet NGEA has a laser ranger in the nose and an improved nav/attack system.

General Dynamics F-16 Fighting Falcon

Type: Single-seat single-engined fighter, developed initially for the air superiority role but since adopted as a swing fighter able to switch from air combat to attack as needed. Variants include fully combat-capable two-seaters with slightly less fuel; austere aircraft; and the two-seat delta-winged F-16F.

When the F-16 Fighting Falcon is mentioned, the usual reaction is to think of it as a fighter with unparalled capability in close combat. It originated with the Lightweight Fighter (LWF) proposal in the early 1970s, won a much publicised flyoff against the Northrop YF-17, and was adopted as an air combat fighter by the USAF and subsequently by many other nations. It has received wide publicity in its designed role, with particular attention being paid to its ability to sustain a 9g turn, although that is possible in only a small section of the flight envelope.

Dimensions	F-16A	F-16C	F-16C MSIP
Length (ft/m)	49.25/15.01	49.25/15.01	49.25/15.01
Span (ft/m)	31.00/9.45	31.00/9.45	31.00/9.45
Height (ft/m)	16.58/5.05	16.58/5.05	16.58/5.05
Wing area (sq ft/m²)	300/27.88	300/27.88	300/27.88
Aspect ratio	3.20	3.20	3.20
Weights			
Empty (lb/kg)	16,234/7,364	17,960/8,150	17,960/8,150
Clean takeoff (lb/kg)	23,810/10,800	26,536/12,040	26,536/12,040
Max takeoff (lb/kg)	35,400/16,057	37,500/17,010	37,500/17,010
Max external load (lb/kg)	15,200/6,895	15,200/6,895	15,200/6,895
Hardpoints	7	7	7
Power	F100-PW-100 tf	F100-PW-100 tf	F100-GE-400 or F100-220
Max (lb st/kN)	23,904/106.3	23,904/106.3	28,000/124.5
Mil (lb st/kN)	14,780/65.7	14,780/65.7	17,000/75.6
Fuel			
Internal (lb/kg)	6,972/3,162	6,972/3,162	6,972/3,162
External (lb/kg)	6,760/3,066	6,760/3,066	6,760/3,066
Fraction	0.29	0.26	0.26
Loadings			
Max thrust	1.00 − 0.68	0.90 − 0.64	1.06 − 0.75
Mil thrust	0.62 − 0.42	0.57 − 0.39	0.64 − 0.45
Wing clean to (lb/sq ft/kg/m²)	79/387	88/432	88/432
Wing max to (lb/sq ft/kg/m²)	118/576	125/610	125/610
Performance			
Vmax hi	M = 2.0	M = 2.0	M = 2.0
Vmax lo	M = 1.2	M = 1.2	M = 1.2
Ceiling (ft/m)	50,000/15,250	50,000/15,250	50,000/15,250
Initial climb (ft/min/m/sec)	50,000/254	50,000/254	50,000/254
Takeoff roll (ft/m)	1,750/533	1,750/533	1,750/533
Landing roll (ft/m)	2,650/808	2,650/808	2,650/808
First flight	20 Jan 1974	19 June 1984	N/A

Its attack capability has received less attention; attacking surface targets does not carry the glamour of air combat, but in fact this role is of hardly less importance to F-16 operators than air superiority. Designed originally as an austere close combat fighter, to add numerical strength to the USAF's relatively small numbers of F-15s, the Fighting Falcon was given considerable ground attack capability at an early development stage. This facility earned it the name of the swing fighter: air superiority was to to be gained by a combination of F-15s and F-16s, whereupon the F-16 could be switched to ground targets. That is an oversimplification: air superiority

can be considerably assisted by attacks on enemy airfields and other air assets on the ground, and the exact point at which the F-16 ceased to operate as an air combat fighter and reverted to the attack role was to be a matter for the commander on the spot.

The F-16 is a small fighter, and for air combat carries an internal 20mm M61 Vulcan cannon with 500 rounds and two Sidewinders mounted on wingtip rails. This basic weapons fit is retained for the air-to-ground mission to give some defensive capability on the outward leg, and turn it into what

Below: Four slicks fall away from an F-16 Fighting Falcon.

amounts to a sweeping fighter on the return trip, needing little or no fighter escort.

Some idea of the importance of the attack mission can be gained from the rather basic APG-66 radar fitted to the F-16A and B, which had no fewer than seven air-to-ground modes. These included real-beam ground mapping, expanded real-beam ground mapping, Doppler beam sharpening, air-to-ground ranging, sea surface search and freeze, the last of which enables the picture on the plan position indicator in the cockpit to remain the same as on the last sweep, showing the aircraft's progress across the screen and allowing the radar to be put on standby until next needed.

The high thrust/weight ratio and moderate wing loading in clean condition allowed an external payload to be carried that was nearly equivalent to the empty weight of the aircraft, while the use of relaxed static stability coupled with fly-by-wire reduced handling problems at high all-up weights and ensured that the pilot could not overstress the aircraft in heavy manoeuvring. Naturally the 9g sustained turn was impossible with the wing pylons laden, and reduced to a limit of 5.5g instantaneous, which is still better than most loaded attack aircraft can manage. The two-seat F-16B has comparable performance to the A except in radius of ac-

tion, the space for the second crew position being achieved at a cost of some 1,187lb (538kg) of fuel.

Inevitably, the F-16 was upgraded, the next variants to appear being the F-16C and D, with the much more capable APG-68 radar and upgraded avionics, which increased the empty weight. A three-phase Multi-Stage Improvement Programme (MSIP) is currently adding an even better avionics fit and more powerful engines, either Pratt & Whitney's F100-220 or General Electric's F110-400, which will restore the F-16C's reduced thrust loadings to better than those of the original F-16A.

These improvements have been expensive, and there has been a move to develop to an austere F-16E with the original APG-66 radar, but apart from the US Navy, which has ordered a modified version with no gun and an F110 engine under the designation F-16N as its new adversary aircraft, there have been no takers.

To date, the F-16 has seen action only with the Israelis, and usually in air combat. The one notable attack mission carried out was the raid on the nuclear reactor at Osirak in Iraq on June 7, 1981, by eight F-16s

Below: An F-16 is seen here equipped with Lantirn pods plus two LGBs under the wings.

escorted by six F-15s. Flying at low level, believed to be about 1,500ft (460m), they traversed first Jordanian and then Saudi Arabian airspace en route to the target from Etzion Air Base in Sinai; each F-16 carried two 2,000lb (907kg) Mk 84 bombs. The primary target was the concrete dome covering the reactor, which was just 105ft (32m) in diameter, and the F-16s attacked in two waves, each consisting of two elements of two aircraft; the first wave scored two direct hits which collapsed the dome onto the reactor, which had not yet been commissioned, while the remaining bombs caused considerable damage to other buildings in the complex. The target having been destroyed, the second wave is reported not to have bombed. Such a high degree of accuracy led many commentators to believe that laser guided bombs had been used, but it later emerged that the Israeli pilots had been practising the mission for over a year.

Two further variants have been evaluated to date, the AFTI (Advanced Fighter Technology Intergration) programme F-16, and the cranked-delta winged F-16XL, now generally called the F-16F. The AFTI F-16 has been used to explore unorthodox

Below: The delta-winged F-16XL demonstrates its impressive bomb-carrying capability.

flight modes, including direct lift, and side force control, which permit de-coupled manoeuvres to be carried out, allowing the nose to be pointed directly at the target while avoiding a direct change of aircraft attitude. For the attack mission such modes should prove extremely valuable in aiming, particularly in a diving attack at low level where it is undesirable to steepen the dive angle further, or where it would be difficult to change the flight direction by orthodox means such as banking.

The F-16F was intended to explore the supersonic cruise and manoeuvre regimes, but its enormous carrying capacity (29 hardpoints), large fuel fraction and semi-recessed low drag carriage of weapons make it a natural choice for the interdiction role. On the other hand, it was rejected by the USAF in favour of the F-15E and has yet to receive an order.

General Dynamics are currently occupied with a proposal to fill the USAF CAS/BAI replacement aircraft, and are rumoured to be working on a dedicated A-16 variant. Some sources have indicated that this will be based on the F-16F, but that seems very unlikely.

Users

Belgium, Denmark, Egypt, Greece, Indonesia, Israel, Netherlands, Norway, Pakistan, South Korea, Thailand, USA, Venezuela

McDonnell Douglas F/A-18 Hornet

Type: Single-seat twin-engined multi-role carrier- and land-based fighter/attack aircraft; fully combat-capable two seaters are designated F-18B. The designation F/A-18 is often used, but is strictly unofficial. A dedicated reconnaissance variant is under development, while the F-18C and D, equipped with ASPJ and compatible with IIR Maverick and Amraam, have entered flight test.

The F/A-18 Hornet marks a departure from the accepted custom of

designing a fighter to fulfil a single role then adapting it for others, being a dedicated multi-role type from the outset. In the early 1970s the US Navy was in the market for a multi-role fighter to replace both its ageing Phantom fighters and its Corsair attack bombers, and attention was focused on the USAF light fighter competition, which was held in the form of a competitive flyoff between the General Dynamics F-16 and the Northrop F-17. It was widely anticipated that the winner of the contest

Dimensions	F/A-18A
Length (ft/m)	56.00/17.07
Span (ft/m)	37.50/11.43
Height (ft/m)	15.29/4.66
Wing area (sq ft/m²)	400/37.17
Aspect ratio	3.52

Weights	
Empty (lb/kg)	21,830/ 9,900
Clean takeoff (lb/kg)	35,800/16,240
Max takeoff (lb/kg)	51,900/23,540
Max external load (lb/kg)	17,000/ 7,711
Hardpoints	5

Power	2 x F404-GE-400 tf
Max (lb st/kN)	16,000/71.2
Mil (lb st/kN)	10,600/47.2

Fuel	
Internal (lb/kg)	10,860/4,925
External (lb/kg)	7,000/3,175
Fraction	0.30

Loadings	
Max thrust	0.89 – 0.62
Mil thrust	0.59 – 0.41
Wing clean to (lb/sq ft/kg/m²)	90/437
Wing max to (lb/sq ft/kg/m²)	130/634

Performance	
Vmax hi	M = 1.8
Vmax lo	M = 1.01
Ceiling (ft/m)	50,000/15,250
Initial climb (ft/min/m/sec)	50,000/254
Takeoff roll (ft/m)	N/A
Landing roll (ft/m)	N/A

First flight	18 Nov 1982

would be selected, but when the Navy came to review its requirements in detail, its decision was that not only did the Northrop entrant provide extra flight safety in the form of two engines, but it also offered more development potential. The flight safety emphasis reflected the fact that most carrier aircraft flights are over the trackless ocean, whereas Air Force missions are often overland.

McDonnell Douglas, with their vast experience in building carrier fighters, teamed with Northrop for the project, which emerged as the F-18, a rather larger aircraft than the F-17 had been and considerably heavier. It was first intended to produce both fighter and attack variants based on a single airframe while Nor-

throp were to build a lighter and slightly more potent export version called the F-18L for land-based use. This was possible because the F-18L did not have to endure the rigours of carrier operations, with catapult launches and arrested landings.

In the event, McDonnell Douglas produced a tour de force. Drawing on the experience of advanced cockpits they had gained on the F-15, they combined HOTAS with CRT multi-function displays to produce what was in effect a new generation cockpit, with few old-fashioned dials and tape instruments. Instead there

Below: Nine Mk 82 bombs fall cleanly away from an early F/A-18 during weapons trials.

were three screens on which information could be called up by computer as required by pushing a few buttons, which enabled a single crewman to command the data needed for either the fighter or the attack mission.

In fighter configuration the Hornet carries two Sidewinders on wingtip rails and two Sparrows conformally on the fuselage; for the attack mission the Sidewinders are retained but the Sparrows are replaced by a FLIR pod and a laser designator and target marker pod in about half an hour. The radar is the Hughes APG-65, with numerous high quality air-to-air and air-to-ground modes, including real beam ground mapping, Doppler beam sharpening sector and patch, terrain avoidance, precision velocity update, and sea surface search, and the comprehensive avionics fit includes RWR, ECM and INS, while computer capacity is almost half as great again as that of the F-15. All in all, the Hornet is a very capable weapon system packaged into an airframe/engine combination which has had few equals.

The Hornet has often been criticis-ed as the loser in the LWF competition, but that is hardly fair. It is far superior as a weapon system to the F-16, and although the Air Force fighter might have a slight edge in close combat it is outclassed at longer ranges; in the attack role there can be little doubt which is the better of the two, a fact which is reflected by the much higher price of the Hornet. It has also been noticeable that wherever the operational requirements have been particularly stringent the Hornet has been selected rather than the competing Fighting Falcon.

The two-seat F-18B is fully combat capable, but the adoption of the second crew position has displaced some 600lb (272kg) of fuel and reduced its operational radius. A reconnaissance variant, the RF-18, is under development, and the upgraded F-18C and two-seat F-18D should enter service before 1990. These will carry the ASPJ and be compatible with Amraam and IIR Maverick. All

Below: A Marine Corps Hornet in attack configuration with Flir and laser designator pods.

variants are armed with a single six-barrel M61 Vulcan cannon with 570 rounds.

A proposed two-seat night and adverse weather attack variant with an advanced nav/attack system and terrain following and synthetic aperture radar, to replace the elderly A-6 Intruder, seems unlikely to proceed.

Users
Australia, Canada, Spain, USA

Above: For the battlefield air interdiction role the Hornet carries cluster munitions.

Below: The Hornet can carry a heavy bomb load as far as the A-7 Corsair, and faster.

Panavia Tornado IDS

Type: Two-seat twin-engined variable-geometry all-weather interdiction/strike aircraft with reconnaissance capability, optimised for the low-level deep penetration blind first pass attack and also used in the anti-shipping role. Variants include a long-range interceptor, Tornado ADV, and an Electronic Combat/Reconnaissance aircraft for the Luftwaffe. A Wild Weasel version has been proposed.

Designed to a joint requirement by the United Kingdom, West Germany and Italy, Tornado IDS is arguably the most effective tactical strike and interdiction aircraft in the world. It is a fairly small aircraft, and in terms of baseline figures is nothing very special. But they are misleading, as its capability to evade defences and hit pinpoint targets at night or in bad weather is unrivalled, a fact amply demonstrated in the USAF Strategic Air Command's annual bombing competitions. Tornados were entered for the first time in 1984, and crews from No 617 Squadron picked up both the Curtis E. LeMay Trophy, awarded to the highest scoring crews in high and low level bombing and time control, and the John C. Meyer Trophy for the highest damage expectancy. They were also second in the LeMay Trophy and second in the Mathis Trophy, for the most points in both high and low level bombing, by very small margins.

Dimensions	Tornado IDS
Length (ft/m)	54.85/16.72
Span (ft/m)	45.58/13.89
Height (ft/m)	19.52/5.95
Wing area (sq ft/m²)	323/30.01
Aspect ratio	6.43 − 2.46

Weights	
Empty (lb/kg)	31,065/14,091
Clean takeoff (lb/kg)	45,000/20,412
Max takeoff (lb/kg)	61,700/27,987
Max external load (lb/kg)	19,800/8,981
Hardpoints	9

Power	2 x RB199 Mk 103 tf
Max (lb st/kN)	15,800/70.2
Mil (lb st/kN)	9,000/40.0

Fuel	
Internal (lb/kg)	11,250/5,100
External (lb/kg)	14,391/6,530
Fraction	0.25

Loadings	
Max thrust	0.70 − 0.51
Mil thrust	0.40 − 0.29
Wing clean to (lb/sq ft/kg/m²)	139/680
Wing max to (lb/sq ft/kg/m²)	191/933

Performance	
Vmax hi	M = 2.2
Vmax lo	M = 1.2
Ceiling (ft/m)	50,000/15,250
Initial climb (ft/min/m/sec)	40,000/203
Takeoff roll (ft/m)	2,500/760
Landing roll (ft/m)	1,600/500

First flight	14 Aug 1974

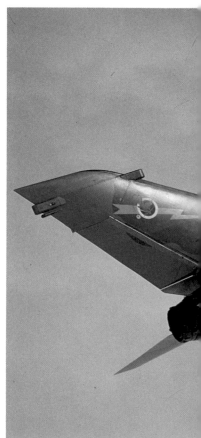

The competition involved three sorties, each of roughly six hours' duration and one of them at night, against simulated SAM defences and interceptors. Roughly 30 weapon releases were made, on which the time error was less than one second, while the average bombing error at low level was less than 60ft (18m). In 1985 crews of No 27 Squadron took part, retaining both the LeMay and Meyer Trophies for the RAF and gaining second place in all three competitions.

The design requirements for Tornado — good short field performance, great lifting capacity and low gust response for high-speed low level flight — were met by using a variable-sweep wing with manual settings of 25°, 45° and 68°. The wing itself probably has more high-lift devices than any other fast jet, with full-span double slotted trailing edge flaps, full-span leading edge slats and Krüger flaps on the fixed wing glove. It therefore lifts out of a short field rather than having to be blasted out with brute-force thrust, while landing takes even less distance: the approach speed is slow, and the bucket-type thrust reversers pull it up sharply once the gear has touched. At maximum sweep gust response is minimised, and Tornado gives a remarkably smooth ride close to the ground.

On its own, high-speed low-level flight is not enough. In clear weather with good visibility it can be achieved manually, but such conditions are

Below: An Italian Tornado IDS of 155 Gruppo/36 Stormo carrying Kormoran anti-ship missiles, a weapon also used by Marineflieger Tornados assigned to maritime strike missions.

also an advantage to the defences. Tornado is equipped to fly at night and in poor weather conditions, which hamper the defences, with terrain-following radar and an incredibly accurate navigation system. The target also has to be found, and attacked accurately, so the terrain-following radar is tied into the autopilot, and keeps the aircraft at a pre-set height above the ground, anywhere between 200ft and 1,500ft (60 — 450m), with three qualities of ride, hard, medium and soft. The combination of height and ride quality selected will be conditioned by the defence capability, and the nature of the terrain.

Tornado will normally be sent against known targets after the mission has been carefully planned and the details entered on a cassette tape, which is inserted into the nav/attack system in the cockpit; waypoints and radar-significant points would be recorded along with other flight data. In flight, INS and Doppler information is constantly fed into the central computer, with an error of less than one mile per hour, and constant updating as radar-significant points are passed gives a very high degree of accuracy. The main computer also knows the speed over the ground very accurately, and can indicate whether the mission is early or late against the planned speed to within a tiny margin. This gives Tornados the capability of flying close formations in conditions where they are unable to see each other; it also allows rapidly sequenced attacks to be carried out from different directions with no danger of a midair collision. Co-ordinated attacks of this nature confuse the ground defences and split their fire, greatly increasing Tornado's survivability.

Tornado's natural targets are airfields, choke points, storage dumps or, in the anti-shipping role, surface vessels, and a wide variety of optimised weaponry can be carried. Two 27mm Mauser cannon are carried as standard, and a pair of Sidewinders can be fitted without wasting a hardpoint, to give some air-to-air capability. RWR, jamming and expendables pods are all normally fitted, while anti-radiation missiles such as Harm or Alarm can be carried.

Various types of powered and unpowered stand-off munitions dispensers are also under development. Such weapons will allow airfields to be attacked without the aircraft having to overfly the target.

Users
Italy, Saudi Arabia, UK, West Germany

Right: Munitions dispensers such as the MW-1 discharging here are an essential part of the Tornado armoury.

Below: Contrails steam from the wingtips as Tornado GR.1s of No. 9 Squadron perform a hard opposition break.

BAe Hawk

Type: Two-seat single-engined advanced trainer developed for clear-weather air defence, close air support, attack and anti-shipping missions. Hawk 200 is a single seat attack/fighter, and the US Navy has adopted the T-45A Goshawk, navalised by McDonnell Douglas, as its new standard advanced trainer.

The mid-1960s requirement for an advanced trainer for the Royal Air Force was originally to have been met by the SEPECAT Jaguar, but development of this aircraft made it rather more capable than had been specified, and consequently rather more expensive. The UK and France then went separate ways, the latter combining with Germany to produce the Alpha Jet while the British requirement resulted in the Hawk.

As was only to be expected, the Alpha Jet and the Hawk became rivals in the export market, with the French trainer, slightly ahead in timing, threatening to scoop the field with the benefit of Dassault's aggressive marketing. The Hawk is very similar in size and general layout, the main difference being that it has only one engine to the Alpha Jet's two, the resulting increased saftey margin apparently being a strong argument

Dimensions	Hawk T.1	Hawk 60	Hawk 200
Length (ft/m)	38.92/11.86	38.92/11.86	37.33/11.38
Span (ft/m)	30.83/9.40	30.83/9.40	30.83/9.40
Height (ft/m)	13.16/4.00	13.16/4.00	13.67/4.17
Wing area (sq ft/m²)	180/16.69	180/16.69	180/16.69
Aspect ratio	5.28	5.28	5.28
Weights			
Empty (lb/kg)	7,450/3,380	8,015/3,635	8,765/3,975
Clean takeoff (lb/kg)	10,700/4,854	11,350/5,148	12,100/5,490
Max takeoff (lb/kg)	12,566/5,700	18,390/8,342	20,065/9,100
Max external load (lb/kg)	5,600/2,540	6,800/3,084	6,800/3,084
Hardpoints	3	5	5
Power	Adour 151	Adour 861	Adour 871
Max (lb st/kN)	N/A	N/A	N/A
Mil (lb st/kN)	5,300/23.6	5,700/23.6	5,845/26.0
Fuel			
Internal (lb/kg)	2,849/1,293	2,927/1,330	2,927/1,330
External (lb/kg)	1,560/708	2,966/1,345	2,966/1,345
Fraction	0.27	0.26	0.24
Loadings			
Max thrust	N/A	N/A	N/A
Mil thrust	0.50 − 0.42	0.50 − 0.31	0.48 − 0.29
Wing clean to (lb/sq ft/kg/m²)	59/290	63/308	67/328
Wing max to (lb/sq ft/kg/m²)	70/341	102/499	111/544
Performance			
Vmax hi	M = 0.98	M = 0.98	M = 0.98
Vmax lo	M = 0.94	M = 0.94	M = 0.94
Ceiling (ft/m)	50,000/15,250	50,000/15,250	50,000/15,250
Initial climb (ft/min/m/sec)	6,000/30.5	9,300/47.0	12,000/61.0
Takeoff roll (ft/m)	2,000/610	2,000/610	2,000/610
Landing roll (ft/m)	2,000/610	1,900/580	1,900/580
First flight	21 Aug 1974	N/A	19 May 1986

in the Alpha Jet's favour, although it was also argued that two engines gave twice as much to go wrong.

As the Hawk matured, it demonstrated that its attrition rate was nothing like what had been projected, that its handling was superior and that its payload/range was far better. The final word can be left to the US Navy; when it evaluated contenders for its advanced trainer requirement it selected the Hawk, subject to its being made carrier-compatible by McDonnell Douglas, refusing to hold a competitive flyoff against the Alpha Jet on the grounds that it would be 'no contest'.

The use of armed trainers in the light attack and close support role is almost obligatory, and it would be a waste of a fast jet not to use it as such. The Hawk therefore underwent development resulting in the export Mk 50, which had improved avionics to the customer's requirements, the Adour Mk 851 engine with the same thrust but better acceleration, greater range, five hardpoints instead of the previous three, and maximum takeoff weight increased to 16,200lb (7,350kg). An optional braking parachute could also be fitted.

The next stage was the Mk 60 series, which featured the more powerful Adour 861, an improved wing, air-to-air missile capability, and a maximum takeoff weight of 18,390lb (8,342kg); a braking parachute was standard on this series. Then came the Hawk 100 advanced ground attack aircraft, with an improved avionics fit, although this never progressed beyond mock-up form and has found no takers.

The failure of the Hawk 100 to find a market can probably be laid at the feet of the Hawk 200, a single-seat dedicated light attack aircraft with vastly increased capability. Various avionics options are available, including an air-to-air radar or a chisel laser nose, while two internal cannon with 150 rounds per gun free the centreline pylon for stores. On previous models it was usual to mount a gun pod in this position. Neither the radar nor the cannon has yet been selected, and the options are open for the customer.

Depending on the exact avionics fit carried, Hawk can operate with a wide variety of stores, including Sea Eagle. It has also proved itself to be no mean performer in the close combat air superiority arena.

Users

Abu Dhabi, Bahrain, Dubai, Finland, Indonesia, Kenya, Kuwait, Saudi Arabia, UK, USA, Zimbabwe

Below: The Hawk 200 is a single-seat light attack variant of the famous two-seat trainer.

Dassault-Breguet Super Etendard

Type: Single-seat single-engined carrier-based strike fighter with limited air-to-air capability.

The Super Etendard began life as a cheap upgrade of the Etendard IVM with improved avionics and a wing modified to give better performance, while the more powerful Atar 8K50 replaced the Atar 8C. In the event,

the Super Etendard emerged as an almost new aeroplane, albeit with a marked family resemblance to the IVM. The main differences are the wing, which has drooping leading edge and double slotted trailing edge flaps, the 8K50 Atar engine, the Thomson CSF/ESD Agave radar optimised for naval missions and a comprehensive nav/attack system.

The official roles of the Super Etendard are given as fleet protection against attack from surface vessels, ground attack, photo-reconnaissance, and fleet air defence, not necessarily in that order. The fleet air defence mission is necessarily of a limited nature; the Super Etendard hardly compares with an F-14, but closely approximates the Sea Harrier in the pursuit and destruction of shadowers such as the Tu-20 Bear, carrying two R550 Magic missiles on the outboard pylons and two 30mm DEFA cannon mounted internally with 125 rounds each.

Super Etendard, with its subsonic performance and unexceptional

Dimensions	Super Etendard
Length (ft/m)	46.96/14.31
Span (ft/m)	31.48/9.59
Height (ft/m)	12.67/3.86
Wing area (sq ft/m²)	306/28.41
Aspect ratio	3.24

Weights	
Empty (lb/kg)	14,330/6,500
Clean takeoff (lb/kg)	20,833/9,450
Max takeoff (lb/kg)	26,455/12,000
Max external load (lb/kg)	4,630/2,100
Hardpoints	5

Power	1 x Atar 9K50 tj
Max (lb st/kN)	N/A
Mil (lb st/kN)	11,025/49.0

Fuel	
Internal (lb/kg)	6,800/3,084
External (lb/kg)	4,800/2,180
Fraction	0.33

Loadings	
Max thrust	N/A
Mil thrust	0.53 − 0.42
Wing clean to (lb/sq ft/kg/m²)	68/333
Wing max to (lb/sq ft/kg/m²)	87/423

Performance	
Vmax hi	M = 1.00
Vmax lo	M = 0.96
Ceiling (ft/m)	45,000/13,700
Initial climb (ft/min/m/sec)	24,600/125
Takeoff roll (ft/m)	N/A
Landing roll (ft/m)	N/A

First flight	28 Oct 1974

appearance, was built in small numbers, some 85 in all, and unlike the majority of Dassault/Breguet aircraft made little impact on the export market. It was one of two non-STOVL aircraft capable of operating from small carriers, the other being the A-4 Skyhawk, which was much cheaper, especially in its refurbished form, while anyone in the market for a land-based attack aircraft had a wide range from which to choose, including attack trainers such as the Hawk and Alpha Jet.

The Super Etendard first achieved prominence in the South Atlantic in 1982 in Argentinian Navy service. A total of four aircraft were serviceable out of five delivered and five Exocets were available: 12 sorties were flown, and all five missiles were launched, resulting in the destruction of HMS *Sheffield* and MV *Atlantic Conveyor*. An element of luck played a part in both these sinkings; *Sheffield* was caught with her defences down, while the *Atlantic Conveyor* was hit by a missile decoyed away from the warships. It is tempting to speculate on the outcome had the old *Ark Royal*, with Gannet AEW aircraft and Phantom fighters, been present.

Super Etendards were next in action over the Lebanon in September 1983, in support of the French contingent of the peace-keeping force; operating from *Foch* and, later, from *Clemenceau*, they provided air support over the subsequent few months.

The next nation to operate the Super Etendard was Iraq, to which a batch of five was supplied, apparently on a sale or return basis pending the delivery of Exocet-compatible Mirage F.1EQ-5s. Details are sparse, but the first operation took place on March 27, 1984. A considerable number of ships have been damaged, but few have sunk and none can definitely be attributed to the Super Etendard. The surviving aircraft (one is believed lost) returned to France early in 1985.

In French service the Super Etendard provides the Aéronavale with a nuclear strike capability carrying the AN-52, which will be replaced by the stand-off ASMP later in the 1980s.

Users
Argentina, France, Iraq

Below: A rocket pod-armed Super Etendard prepares to launch from a French aircraft carrier. The type will also carry the ASMP nuclear missile.

SOKO J-22 Orao/CNIAR IAR 93

Type: Single-seat twin-engined attack aircraft with two-seat combat-capable training version.

The Orao/IAR-93 is a collaborative programme between Yugoslavia (Orao) and Romania (IAR-93), and is remarkable for the extreme lengths to which the partners have gone to avoid upstaging each other; both the initial prototypes, one made in each country, and the initial two-seat versions first flew on the same day.

Orao/IAR-93 was first produced in an A variant, powered by two Rolls-Royce Viper turbojets, reliable engines that had been used in Yugoslavia for many years. Little has been released about the avionics fit, but this is believed to consist of little more than communications and adverse-weather flight instrumentation.

The type's primary mission is close air support, with low and medium air superiority and point defence an added capability. The thirsty turbojets and fairly modest fuel fraction do not

Dimensions	OraoIAR-93B
Length (ft/m)	45.93/14.00
Span (ft/m)	31.56/9.62
Height (ft/m)	14.60/4.45
Wing area (sq ft/m²)	280/26.00
Aspect ratio	3.56

Weights	
Empty (lb/kg)	12,566/5,700
Clean takeoff (lb/kg)	17,500/7,938
Max takeoff (lb/kg)	23,150/10,500
Max external load (lb/kg)	6,173/2,800
Hardpoints	5

Power	2 x RR Viper 633 tj
Max (lb st/kN)	5,000/22.2
Mil (lb st/kN)	4,200/18.9

Fuel	
Internal (lb/kg)	4,465/2,025
External (lb/kg)	2,787/1,264
Fraction	0.26

Loadings	
Max thrust	0.57 – 0.43
Mil thrust	0.46 – 0.35
Wing clean to (lb/sq ft/kg/m²)	63/305
Wing max to (lb/sq ft/kg/m²)	83/404

Performance	
Vmax hi	M = 0.92
Vmax lo	M = 0.95
Ceiling (ft/m)	42,600/13,000
Initial climb (ft/min/m/sec)	12,992/66
Takeoff roll (ft/m)	2,260/690
Landing roll (ft/m)	3,450/1,050

First flight	21 Oct 1974

give a wide radius of action, but in the CAS mission this is not too important. Two 23mm GSh-23 twin barrel cannon are fitted internally for strafing or air-to-air use, with a capacity of 200 rounds per gun.

The B variant, which first flew towards the end of 1983, has afterburners fitted to its Viper 663 engines, which has improved short takeoff performance and almost doubled the initial rate of climb, as well as presumably making handling more sprightly, although it has failed to provide supersonic performance. No details are forthcoming about the avionics, but it may be assumed that it has a modern nav/attack system and laser ranging may be fitted. Normal weapons load is about half the stated maximum, while the operational radius using a hi-lo-hi profile is about 194nm (370km).

Users
Romania, Yugoslavia

Below: The light attack field is one in which many countries have experimented, and the SOKO Orao, seen here armed with a rocket pod, is no more than an average product.

Bottom: The Orao/IAR-93 design was conditioned by the engines available, in this case Rolls-Royce Vipers. Even with reheat it is firmly subsonic.

Aermacchi MB-339K Veltro 2 and MB-339C

Type: Two-seat single-engined attack/trainer and single-seat single-engined light attack fighter. Other variants are two-seat trainers with some attack capability.

The MB.339 is essentially a development of the MB.326 jet trainer first flown in December 1957 which has been used by many countries and has been licence-built in Australia, Brazil (as the Xavante) and South Africa (as the Impala). A single-seat light attack variant was also produced as the MB.326K. While following largely the same lines as the 326, the MB.339 has a redesigned structure with a raised second seat to improve the instructor's forward view. Powered by an unreheated Rolls-Royce Viper, its performance is unexceptional, although handling is reputed to be very crisp and precise, and the MB.339PAN variant is used by the Frecce Tricolori, the Italian Air Force aerobatic team. Compared with, for

Dimensions	MB.339C	MB.339K Veltro 2
Length (ft/m)	36.00/10.97	36.09/11.00
Span (ft/m)	35.63/10.86	36.75/11.20
Height (ft/m)	13.10/3.99	13.10/3.99
Wing area (sq ft/m²)	208/19.30	208/19.30
Aspect ratio	6.10	6.49
Weights		
Empty (lb/kg)	6,890/3,125	7,154/3,245
Clean takeoff (lb/kg)	9,700/4,400	11,000/4,990
Max takeoff (lb/kg)	13,000/5,900	14,000/6,350
Max external load (lb/kg)	4,000/1,815	4,000/1,815
Hardpoints	6	6
Power	1 x Viper 632-43 tj	1 x Viper 680-43 tj
Max (lb st/kN)	N/A	N/A
Mil (lb st/kN)	4,000/17.8	4,450/19.8
Fuel		
Internal (lb/kg)	2,421/1,098	3,486/1,580
External (lb/kg)	N/A	N/A
Fraction	0.25	0.32
Loadings		
Max thrust	N/A	N/A
Mil thrust	0.41 − 0.31	0.40 0.32
Wing clean to (lb/sq ft/kg/m²)	47/228	53/259
Wing max to (lb/sq ft/kg/m²)	63/306	67/329
Performance		
Vmax hi	M = 0.77	M = 0.77
Vmax lo	M = 0.85	M = 0.85
Ceiling (ft/m)	47,900/14,600	46,000/14,000
Initial climb (ft/min/m/sec)	6,595/33.5	10,000/51
Takeoff roll (ft/m)	1,525/465	1,900/580
Landing roll (ft/m)	1,360/415	1,475/450
First flight	12 Aug 1976 (prototype)	N/A

example, the BAe Hawk, the MB.339 is short on range and carrying capacity as well as performance, but is still good value for money, and at a presentation in Zurich in 1978 it was demonstrated to be superior in terms of warload/attrition costs, given a standard rate of attrition. On the other hand, against modern ground defences it is less effective and less survivable.

Five MB.339s of the Argentine Navy took part in the South Atlantic conflict in 1982, operating from Port Stanley, and one of them delivered the first attack on the landing force in

Below: Gun pods and bombs on an aerobatic MB-339PAN.

San Carlos Water, strafing HMS *Argonaut* with cannon and rocket fire and inflicting minor damage. Otherwise they seem to have done little or no damage: one crashed in bad weather on May 3, another was shot down by a Blowpipe missile over Goose Green on May 28, and the remaining three were all captured in various states of disrepair at Port Stanley.

Following the example of the MB.326K, Aermacchi developed a single-seat variant, the MB.339K Veltro II, in 1980. Unlike the two-seaters, which have no internal cannon but can carry 30mm DEFA pods under the wings, the Veltro has two internal 30mm cannon with 125 rounds each, and for the anti-helicopter mission it can carry two 30mm gun pods, giving a heavy rate of fire. Equipped with a digital nav/attack system, HUD, Doppler INS and a stores management system, it is also being cleared to carry two Sidewinders in addition to a wide variety of other stores, though these are limited by the lack of ground clearance and hardpoint weight limitations. Veltro II is also offered in the anti-ship role, with the Marte Mk 2 missile and a compatible radar. Unlike the MB.326K, the Veltro has not sold, the initial order for Peru having been cancelled.

Latest in the line is the MB.339C, a two-seat trainer first flown in December 1985 which is offered with the same avionics package as Veltro II and is optimised for close air support, anti-helicopter missions and the ASV role. It also carries the ELT-156 RWS, active ECM, and countermeasurers expendables, while proposed armament includes Sidewinder and AGM-65 Maverick, plus Marte complete with radar pod as on the Veltro.

The systems upgrading envisaged carries its own penalty, that of increased cost: every improvement brings costs nearer to those of a more capable aircraft such as the Hawk.

Users
Argentina, Dubai, Italy, Malaysia, Nigeria, Peru

Left: The civil registration on the private-venture prototype MB-339K Veltro 2 is apt.

Sukhoi Su-25 Frogfoot

Type: Single-seat twin-engined close air support aircraft. Rumours that a two-seater has been built have yet to be confirmed; there are reports of simulated carrier operations.

In many respects the Su-25 Frogfoot is an enigma to the West. In some ways it appears to be a Soviet equivalent of the Fairchild A-10, while bearing a vague resemblance to the losing competitor in the A-X programme, the Northrop A-9; the dimensions are known only approximately; and the weights are very much a matter of guesswork, with no

Dimensions	Su-25 Frogfoot
Length (ft/m)	49.22/15.00
Span (ft/m)	46.90/14.30
Height (ft/m)	16.41/5.00
Wing area (sq ft/m²)	420/39.00
Aspect ratio	5.24

Weights	
Empty (lb/kg)	19,200/8,709
Clean takeoff (lb/kg)	28,000/12,700
Max takeoff (lb/kg)	40,000/18,144
Max external load (lb/kg)	12,000/5,443
Hardpoints	10

Power	2 x R13-300 tj
Max (lb st/kN)	N/A
Mil (lb st/kN)	11,250/50.0

Fuel	
Internal (lb/kg)	8,400/3,810
External (lb/kg)	2,060/934
Fraction	0.30

Loadings	
Max thrust	N/A
Mil thrust	0.80 − 0.56
Wing clean to (lb/sq ft/kg/m²)	67/327
Wing max to (lb/sq ft/kg/m²)	95/464

Performance	
Vmax hi	N/A
Vmax lo	M = 0.74
Ceiling (ft/m)	N/A
Initial climb (ft/min/m/sec)	N/A
Takeoff roll (ft/m)	1,500/457
Landing roll (ft/m)	1,200/366

First flight	1977

two sources seeming to guess the same. All that can be said with any certainty is that the Frogfoot is smaller, lighter and faster than the A-10. The American DoD credits it with an operational radius of about 300nm (556km), presumably with a standard warload, which is probably about half the maximum; the mission profile is not given, but can be assumed to be at medium to low level. That is considerably less than the A-10 can achieve, although radius of action is basically irrelevant for both types, loiter time being more important, while penetration of more than a few miles into defended airspace is virtual suicide.

It has been widely speculated that the Su-25 is a continuation of the World War II Shturmovik tradition, although for all practical purposes this had fallen into disuse over the succeeding three decades. There does, however, seem to be a trend that where the United States leads, the Soviet Union will follow, the classic example of which is the simarity of the Su-24 Fencer to the F-111. It may well be that the Soviets liked the idea of a survivable close air support machine along the lines of the A-10, even though they chose to make it a bit faster. The Frogfoot is also certain to be heavily armoured; the pilot appears to be well protected and, in contrast to his American counterpart, sits well back in the cockpit transparency, where he appears to be well masked from fire from the rear quarter. This position is achieved at the cost of almost non-existent rear vision, three mirrors mounted on the canopy bow being the only way of spotting a rear quarter threat.

The cockpit is probably well protected, although it is doubtful whether the Soviet designers have come up with a titanium bathtub; steel is more likely. The weights given have been calculated from various assumptions, but the empty weight still looks rather high for the size when the A-10 figures are considered, but armour to the engines, underside and rear of the aircraft would account for that, as well as being a traditional hallmark of the Shturmovik.

The Su-25 appears to have a twin-barrel 30mm cannon mounted under the cockpit floor, with the barrels off-set to the left. While this might just be an anti-armour weapon, it seems very unlikely that it has anything like the hitting power of the A-10's GAU-8. The nose has a chisel shape generally associated with a laser-ranger target spot marker.

To date, Frogfoot has only seen in Afghanistan, where it has been extensively used against the Mujahi-deen; it has performed very well in the counter-insurgency role, usually operating in pairs, with one flying at very low level while the higher aircraft dispenses IR decoy flares to attract heat-seeking SAMs fired from shoulder-mounted launchers. It has also been reported to operate in conjunction with Mi-26 Hind attack helicopters, which would accord with Soviet battlefield practice.

Users
Czechoslovakia, Hungary, Iraq, USSR

Above: Often assumed to be a mirror image of the A-10, Frogfoot resembles the A-9.

Below: The four main pylons under each wing are the principal warload carriers.

Dassault-Breguet Mirage 2000N

Type: Two-seat single-engined nuclear strike/attack aircraft. Other variants are the 2000C single-seat interceptor/air superiority fighter with multi-role capability, including air-to-ground; the 2000B two-seat conversion trainer; and the 2000R reconnaissance aircraft. Export models have other designations.

The Mirage 2000 resulted from an Armée de l'Air requirement for a multi-role fighter capable of intercepting very high-speed, high-altitude intruders; it was also to replace the elderly Mirage III and the Mirage F.1. The maximum speed of Mach 3 originally specified was gradually lowered to Mach 2.7, and then still further reduced. Two factors influenced the relaxation of the top speed requirement; cost was certainly one, as the proposed new fighter would have been some two and a half times as expensive as the Mirage F.1 to procure; the other was the projected performance of the next generation of missiles.

Dimensions	Mirage 2000N
Length (ft/m)	46.50/14.17
Span (ft/m)	29.50/8.99
Height (ft/m)	N/A
Wing area (sq ft/m²)	441/40.98
Aspect ratio	1.97

Weights	
Empty (lb/kg)	17,000/7,710
Clean takeoff (lb/kg)	23,750/10,773
Max takeoff (lb/kg)	36,375/16,500
Max external load (lb/kg)	16,755/7,600
Hardpoints	9

Power	1 x SNECMA M53-P2 tf
Max (lb st/kN)	21,400/95.0
Mil (lb st/kN)	14,400/64.0

Fuel	
Internal (lb/kg)	6,346/2,880
External (lb/kg)	8,758/3,973
Fraction	0.27

Loadings	
Max thrust	0.90 – 0.59
Mil thrust	0.61 – 0.40
Wing clean to (lb/sq ft/kg/m²)	54/263
Wing max to (lb/sq ft/kg/m²)	82/403

Performance	
Vmax hi	M = 2.35
Vmax lo	M = 1.20
Ceiling (ft/m)	60,000/18,300
Initial climb (ft/min/m/sec)	49,212/250
Takeoff roll (ft/m)	N/A
Landing roll (ft/m)	1,200/410

First flight	20 Nov 1982

At the time that the Super Mirage was cancelled, Dassault were working on a simpler and cheaper aircraft, the Mirage 2000, a reversion to the simple delta wing of the Mirage III, which had been superseded by the more orthodox tailed Mirage F.1. The delta layout had been well suited to high-speed, high-altitude flight, but had certain disadvantages in manoeuvre combat. However, the advent of relaxed static stability combined with quadruplex fly-by-wire could produce a very manoeuvrable fighter which retained the high-speed, high-altitude advantages of the delta planform. Variable camber also helped, the full-span leading edge slats operating automatically as a function of angle of attack when the undercarriage was up, and combining with two-section elevons to the entire trailing edge. Small strakes were fitted to the sides of the engine inlets to produce a vortex, and to reduce the download. And to offset the limited power available, composite materials were used extensively to reduce weight.

The result was a small, fast and highly manoeuvrable fighter with a good rate of climb, and one which was affordable in the sort of quantities that made sense, making it an attractive proposition in the export market. For the attack role a total of

Below: Eighteen BAP 100 anti-runway bombs on the centreline of a Mirage 2000.

nine hardpoints were built in, although two of these have a limit of 660lb (300kg) and four more are limited to 880lb (400kg), leaving only three with a heavy load capability — some 3,970lb (1,800kg). While other stores can be carried, the lighter rated hardpoints are generally used for AAMs or equipment pods, depending on the needs of the specific mission. Two 30mm DEFA 554 cannon with 125 rounds each are installed internally.

The Thomson-CSF RDM multimode radar has, in addition to its air-to-air modes, certain air-to-ground functions, namely ground mapping, contour mapping, terrain avoidance, air-to-ground ranging and sea search and track; coupled with a state of the art nav/attack system and an internal ECM suite, it provides a very fair air-to-ground capability.

The Mirage 2000N has a two-man crew and is based on the 2000B two seat conversion trainer. In place of the RDF radar, it has the ESD/TH-CSF Antilope radar with a terrain-following mode, optimised for air-to-ground weapons delivery and deep penetration at low level in darkness or adverse weather. This allows automatic flight at 300ft (90m) above ground level, to be reduced to 200ft (60m) in future, and is coupled with a SAGEM INS and a comprehensive avionics suite which allows first-pass blind strikes to be made on pre-targeted points. While such a capability could have been built into a single-seat aircraft, pilot workload would have been too high, so a second crew member was given responsibility for navigation, monitoring the radar, operating the ECM systems and weapons management.

In the nuclear strike role the main weapon will be the ASMP stand-off missile, which has a 150kT warhead, carried on the centreline. The second crew position could only be squeezed in at the expense of some 700lb (318kg) of fuel, and since penetration will in most cases have to be made at low level this considerably reduces the operational radius. The Mirage 2000 is equipped for in-flight refuelling, but that is hardly a practical proposition anywhere near hostile territory, so drop tanks will be carried on the two inboard wing pylons, while AAMs, jamming pods and ECM expendables pods will probably occupy the remaining hardpoints. No reliable figures are available as to payload/range, but even with two tanks carried externally the Mirage 2000N can hardly be regarded as a strategic weapon.

Users
Abu Dhabi, Egypt, France, India, Greece, Peru

Right: The Mirage 2000's delta configuration gives lots of room for stores: this example carries BM 400 area saturation bombs.

Below: Eight 250kg bombs plus tanks and two Magic missiles for self defence form another possible weapons load.

McDonnell Douglas/BAe Harrier II

Type: Single-seat single-engined attack and reconnaissance fighter with vertical/short takeoff and landing capability used by the US Marine Corps in the close air support role and by the RAF as a Harrier GR.3 replacement.

The Harrier II is a straight development from the earlier Harrier GR.3/AV-8A and is intended to better the earlier model in everything except maximum speed. It remains obviously a Harrier, but with certain external differences: the cockpit, which is closely faired into the fuselage on the Harrier I, has been raised by about 10in (25cm) to give a better view forward, down and to the rear; the wing is greater in area and of increased span, with a consequent improvement in both wing and span loading, and is of a different, supercritical section — one consequence of which is that it can carry a considerable amount of extra fuel — with large, positive circulation flaps to the trailing edge; the outriggers have been

Dimensions	Harrier II
Length (ft/m)	46.33/14.12
Span (ft/m)	30.33/9.24
Height (ft/m)	11.65/3.55
Wing area (sq ft/m²)	230/21.37
Aspect ratio	4.00

Weights	
Empty (lb/kg)	12,922/5,861
Clean takeoff (lb/kg)	20,386/9,247
Max takeoff (lb/kg)	29,750/13,495
Max external load (lb/kg)	9,200/4,173
Hardpoints	7

Power	1 x Pegasus Mk 105/F402-RR-406 tf
Max (lb st/kN)	N/A
Mil (lb st/kN)	21,700/96.4

Fuel	
Internal (lb/kg)	7,142/3,240
External (lb/kg)	7,800/3,538
Fraction	0.35

Loadings	
Max thrust	N/A
Mil thrust	1.06 − 0.73
Wing clean to (lb/sq ft/kg/m²)	89/433
Wing max to (lb/sq ft/kg/m²)	129/632

Performance	
Vmax hi	M = 0.93
Vmax lo	M = 0.88
Ceiling (ft/m)	45,000/13,700 +
Initial climb (ft/min/m/sec)	Not released
Takeoff roll (ft/m)	1,000/300
Landing roll (ft/m)	Vertical

First flight	9 Nov 1978

brought inboard; and a leading edge root extension has been added to give a positive destabilisation effect for added manoeuvrability. The last feature also provides a vortex to clean up the boundary layer flow. The engine inlets have a more elliptical shape as well as a single row of auxiliary doors, while lift improvement devices (LIDs) have been installed under the fuselage to control the efflux circulation during vertical takeoff and landing.

Changes below the surface have been just as far reaching. Composite materials are used extensively to save weight; the avionics have been substantially improved, though the fit varies between the AV-8B and the GR.5, and a new stability augmenta-tion system has made a tremendous improvement to flying qualities in the hover. The Harrier I was always slightly unstable in hovering flight and could be distinctly twitchy at certain angles of bank, but the Harrier II is much more solid and has been described as running as though on rails, which is very helpful at night and in poor weather as well as reducing pilot workload, which in Harrier I is acknowledged to be high during the landing phase.

Weapons delivery is effected by the Hughes ARBS (Angle Rate Bombing System), which combines an

Below: An AV-8B Harrier II of the USMC carries a total of seven Snakeye retarded bombs.

electro-optical sensor and a televisual contrast tracker in the nose and can be used for a variety of weapons, including AGM-65 Maverick; auto-release or depressed sight-line attacks are avaliable. Both American and British aircraft carry internal ECM suites and expendables dispensers, and the avionics fit differs mainly in detail.

Differences between the AV-8B and the GR.5 are as follows: the GR.5 has better hardening against bird strikes, the nose cone, wing leading edges and intake lips have all been strengthened, and the windshield is 50 per cent thicker; the GR.5 carries two 25mm Aden cannon in streamlined pods under the fuselage whereas the AV-8B has a single five-barrel GAU-12 25mm rotary cannon on one side and an ammunition tank on the other; the AV-8B has a Stencel ejection seat while the GR.5 has a Martin-Baker Mk 10; and the GR.5 has a moving map display as fitted in the GR.3 while the AV-8B carries the ASN-130 INS, although it may also carry a moving map display in the future.

Moves are afoot to give the Harrier II a night attack capability, probably FLIR for the GR.5 and Lantirn for the AV-8B. This seems a little strange, because GR.3s from Wittering have been practising night attacks on the range at Sutton Bridge for many years and on quite dark nights. However, the ability to perform their low level routine around the clock would be a great advantage.

The final difference between the AV-8B and the GR.5 is that the latter is to have two extra hardpoints under the wings and in line with the outriggers. These are dedicated to Sidewinder AAMs for self defence, enabling them to be carried without penalising the main hardpoints; in all it is possible for the GR.5 to carry six Sidewinders against the four of the AV-8B. The combination of Sidewinders and gun makes Harrier II a formidable adversary in close combat, although without radar it is disadvantaged before the merge and reliant on the RWR and/or visual sighting.

Clever weapon aiming systems apart, Harrier II far surpasses the Harrier I in payload/range, which is ef-

fectively doubled, allowing Harrier II to carry either double the load for the same distance, or the same load twice as far, thus silencing critics of the earlier aircraft. The Harrier is rarely if ever going to be used on a deep penetration sortie, so its operational effectiveness will lie in the increased hitting power. Actual figures are hard to come by, but in 1983 an AV-8B flew a demonstration hi-lo-hi profile while carrying seven 570lb (250kg) bombs, depositing them on target at a distance of 422nm (782kg).

Users
Spain, UK, USA

Right: An AV-8B drops a stick of Mk 82 slicks in level flight during weapons trials.

Below: The maximum load of the AV-8B is shown here: a total of 15 Mk 82 500lb bombs.

Aeritalia/Aermacchi/ EMBRAER AMX

Type: Single-seat single-engined multi-role aircraft with the accent on attack but with some air-to-air and reconnaissance capability. Two-seat variants are proposed for advanced training, anti-shipping, and electronic warfare.

Design studies for AMX began in 1977, the objective being an aircraft to replace the Fiat G.91 in the light attack and reconnaissance roles and the·F-104 in both those roles plus counter-air and anti-shipping attack.

A consortium set up between Aeritalia and Aermacchi was later joined by Embraer of Brazil, whose air arm wanted a comparable machine. Although firm orders have been placed by both Italy and Brazil, and other countries have been reported as showing interest, AMX appears to be ploughing a lonely furrow. Nowhere else has a firmly subsonic dedicated attack machine been developed from scratch, nor are AMX's baseline performance figures anything startling. In the international marketplace it is

Dimensions	AMX
Length (ft/m)	44.52/13.57
Span (ft/m)	29.10/8.87
Height (ft/m)	14.99/4.57
Wing area (sq ft/m²)	226/21.00
Aspect ratio	3.75

Weights	
Empty (lb/kg)	14,771/6,700
Clean takeoff (lb/kg)	21,164/9,600
Max takeoff (lb/kg)	27,558/12,500
Max external load (lb/kg)	8,377/3,800
Hardpoints	5

Power	1 x Spey RB168-807 tf
Max (lb st/kN)	N/A
Mil (lb st/kN)	11,000/49.0

Fuel	
Internal (lb/kg)	6,000/2,720
External (lb/kg)	N/A
Fraction	0.28

Loadings	
Max thrust	N/A
Mil thrust	0.52 − 0.40
Wing clean to (lb/sq ft/kg/m²)	94/457
Wing max to (lb/sq ft/kg/m²)	122/595

Performance	
Vmax hi	N/A
Vmax lo	M = 0.86
Ceiling (ft/m)	42,650/13,000
Initial climb (ft/min/m/sec)	Not released
Takeoff roll (ft/m)	2,460/750
Landing roll (ft/m)	N/A

First flight	May 1983

competing both with refurbished old designs upgraded with advanced avionics fits and with adapted and armed advanced trainers, both of which are going to work out much cheaper than AMX, which is quite a large aircraft for its payload/range performance.

Little firm information has been released about the avionics fit, but it is known that an advanced nav/attack system has been tailored around two main computers via a digital data bus, while plenty of space exists to allow extra systems to be fitted as necessary; an internal ECM suite is carried, incorporating RWR and both active jammers and expendables. For reconnaissance missions pallet-mounted photographic systems can be loaded, or an infra red/optronics pod can be hung on the centreline. The Brazilian and Italian aircraft differ in weaponry: the latter have a single 20mm M61 Vulcan cannon mounted internally, whereas the Brazilians have been forced by US export restrictions to mount two 30mm DEFA cannon.

AMX (the name has yet to be decided but it has to mean the same in both Italian and Portuguese) has been designed to haul a moderate load out of a short or semi-prepared airfield, and take it a moderate distance at high subsonic speed and

Below: The AMX — a name has yet to be chosen — appears to be an unexceptional aircraft.

125

low level, by day or by night, and in less than perfect weather conditions. Forecasts have been made that there is a large market for the anti-shipping variant "with practically no opposition". If that means there is little opposition from purpose-made aircraft, the statement is correct, but in practice there are many fighters and even more armed trainers that can carry specialised anti-ship missiles such as Exocet, Sea Eagle, Kormoran or Harpoon. In fact, it would be a singularly small or ill-equipped aircraft that could not be fitted out to both carry and launch at least one of these weapons.

Comparisons are odious, but difficult to avoid in the case of AMX and Hawk 200. The British aircraft carries 18 per cent less rather faster, and using a hi-lo-lo-hi profile, for considerably further than the AMX; using a lo-lo profile the gap is substantially reduced but still exists. Hawk 200 has better short-field performance, and is far superior in air combat should it have to fight its way home, and while the licence-built Spey in the AMX gives considerably more thrust, the heavier weight of the bigger machine makes the difference between the respective thrust loadings marginal. At the bottom line weight equates roughly with cost, and AMX is some 6,000lb (2,725kg) or 69 per cent heavier than the Hawk, which, on a strict knock for knock basis, would mean five Hawk 200s for the price of three AMXs. An even more attrractive option would be to acquire the Hawk 60 series and add advanced training to the list of roles.

On the other hand, AMX is large enough to accommodate considerable growth in terms of avionics systems and has been designed with a surplus of power in both electrical and hydraulic systems, although the two-seat version will lose one fuselage fuel tank, which will reduce its range and endurance. At the same time, any great increase in weight will have an adverse effect on the thrust/weight ratio and increase the already rather high wing loading, with a consequent diminution of short field performance and acceleration. The rather small wing is already well equipped with high-lift devices, with full-span leading edge slats and large double slotted trailing edge flaps. BLC blowing could be used to create more lift for takeoff and landing, but only at the expense of reducing the thrust available.

Carrying an external load of 6,000lb (2,720kg), with a gross takeoff weight, AMX has an attack radius of 280nm (520km) using a hi-lo-lo-hi profile, while in the lo-lo mission at the same weights the figures 200nm (370km).

Against the cheaper alternatives of refurbished attack fighters and armed trainers AMX's chances in the export market do not look good. Argentina has expressed an interest, but the need to provide an alternative engine poses great difficulty, as Britain is hardly likely to permit export of the Spey to that country. On the other hand, the AMX may yet surprise us all.

Users
Brazil, Italy

Above: The Brazilian AMX has two 30mm DEFA cannon; the Italian model has a single M61.

Below: An Italian AMX displays its four underwing and single fuselage hardpoints.

Israel Aircraft Industries Lavi

Type: Single-seat single-engined attack and and close air support fighter, with considerable secondary capability in the air superiority role, and two-seat, fully combat-capable conversion trainer.

In the late 1970s work was started on a project called Arieh, which was intended to be a replacement for the A-4 Skyhawk in Israeli service: heavily dependent on American technical cooperation, it was dropped when it became obvious that export would

Dimensions	Lavi
Length (ft/m)	47.21/14.39
Span (ft/m)	28.61/8.72
Height (ft/m)	15.68/4.78
Wing area (sq ft/m²)	414/30.50
Aspect ratio	2.10

Weights	
Empty (lb/kg)	17,000/7,711
Clean takeoff (lb/kg)	23,500/10,660
Max takeoff (lb/kg)	37,500/17,000
Max external load (lb/kg)	15,400/6,985
Hardpoints	14+

Power	1 x PW1120 tf
Max (lb st/kN)	20,620/91.6
Mil (lb st/kN)	13,550/60.2

Fuel	
Internal (lb/kg)	6,000/2,722
External (lb/kg)	N/A
Fraction	0.26

Loadings	
Max thrust	0.88 − 0.55
Mil thrust	0.58 − 0.36
Wing clean to (lb/sq ft/kg/m²)	57/277
Wing max to (lb/sq ft/kg/m²)	91/442

Performance	
Vmax hi	M = 1.85
Vmax lo	M = 1.05
Ceiling (ft/m)	N/A
Initial climb (ft/min/m/sec)	N/A
Takeoff roll (ft/m)	N/A
Landing roll (ft/m)	N/A

First flight	31 Dec 1986

be difficult if not impossible as a direct consequence of the American technology incorporated.

Later Israeli requirements were for a larger and more capable aircraft optimised for interdiction and close air support and, certain American objections having been overcome or lifted, the project went ahead as the Lavi. Air defence and air superiority were listed as secondary roles, but in fact the Israeli machine seems to have been given an air-to-air capability far beyond that of the usual dedicated attack fighter, and in appearance looks more like an air superiority fighter with ground attack capability, in that it has followed the latest trend of Western fighters in having a delta wing with movable canard foreplanes, relaxed static stability, and quadruplex fly-by-wire.

Lavi is to have an Elta multi-mode radar, an advanced nav/attack system, and an internal ECM suite. A chin intake similar to that of the F-16 has been adopted, and two ventral fins have appeared, which seems to indicate that a fair amount of high angle of attack work is envisaged. Predicted turn rates are 13.2°/sec sustained and 24.3°/sec instantaneous, figures more reminisent of an air superiority fighter than an attack type, while Sidewinder, Shafrir or Python missiles will be carried as standard on wingtip rails, and a 30mm DEFA cannon is fitted internally.

Lavi will have a state of the art cockpit, with a wide-angle diffractive optics HUD and three multi-function CRT displays which give clear presentation even in bright sunlight. The Israeli Air Force operates the F-16, and experience with this aircraft has led it to abandon the reclining seat and sidestick controller in favour of the orthodox upright seat and central control column in the Lavi. The F-16's inclined seat had been found to induce stiff necks and shoulder strains in high-g manoeuvring, while the sidestick used valuable console space and made it impossible for a pilot with a disabling wound in

the right arm to recover to base using only his left.

External weapons appear to be carried semi-conformally, recessed into the underside, with three points to each side of the fuselage ranged front to back, plus two centreline points; three more stations under each wing make up the number. Accuracy in aiming at ground targets will be assisted by direct lift control, made possible by the nine computer-controlled movable surfaces. Side force control has not been mentioned, but doubtless has been examined.

Performance reports, like everything else, are conflicting — the tabular data is provisional for everything but the engines and range assessment is difficult. However, IAI have stated that using a hi-lo-hi profile

Top: The Lavi configuration is more reminiscent of a fighter than an attack aircraft.

Above: The first Lavi flew on the last day of 1986.

gives Lavi a mission radius of more than 1,000nm (1,853km) with a payload of six Mk 82 or two Mk 84 bombs. Frankly, this seems a little ambitious.

By late 1986 it was doubtful whether Lavi would proceed: the design is heavily dependent on US technology and funding, and it is being argued that it is too costly and will do nothing that the F-16 cannot do.

User
Israel (possibly)

Combat Tactics

THE task of the attack aircraft can be defined as to hit the target swiftly and accurately with whatever munitions are carried, and return safely to base. That is not to imply that avoiding losses is the major factor — war is war, and losses will occur — but keeping the attrition level within tolerable limits remains one of the priorities. To carry out the task as stated three conditions must be satisfied: the enemy defences must be penetrated, both outbound and on the return leg; the target must be located and correctly identified; and aiming has to be accurate.

Methods of penetrating enemy defences vary according to the strength and sophistication of the hostile detection, reporting, command and control network, and how much intelligence is available about its capabilities. The usual worst case yardstick is generally held to be an all-out conventional war in central Europe, but it should be remembered that the central European scenario is, barring grave political miscalculations, an unlikely one, while limited wars in other parts of the globe occur quite frequently.

No country in the world can muster a comprehensive, multi-layered air defence. There will always be strong points, usually centred on key features such as airfields, ground radars and command posts, while communications bottlenecks such as bridges and roads through constricting geographical features will normally have better than average defences against air attack. By the same token, there will be areas where the counter-air assets are spread thinly, and where there are gaps or shadows in the radar coverage. Careful planning can route the attack force around the strongly defended areas wherever possible and make the best use of deceptive measures, including flight paths chosen to keep the enemy guessing as to the identify of the actual target until the last possible moment, though radical changes of course will make fuel state more critical and in-

Route planning

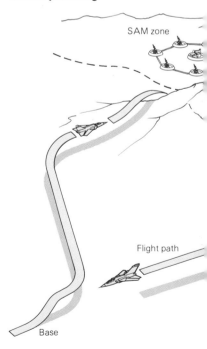

SAM zone

Flight path

Base

Below: Peacetime safety and noise rules prevent these bomb-armed RAF Tornados from flying at operational heights.

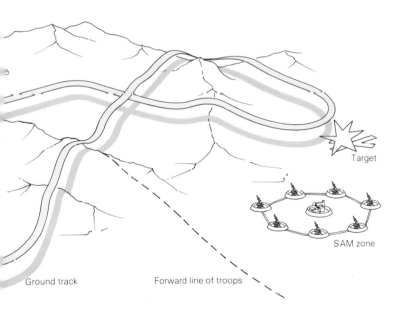

Target

SAM zone

Ground track

Forward line of troops

In a typical low-level strike mission aircrew will attempt to use terrain features to conceal their approach to the target and their return flight to base. The SAM site near the target is thus partially countered, but the more distant site is still a threat.

crease the time spent at risk over hostile territory. Mission planning will also involve striking a balance between fuel and munitions in determining the load to be carried.

The mission profile chosen for a specific strike is an important factor. At high altitudes either maximum speed or fuel economy at cruise speeds will be improved, but the distance at which the aircraft can be detected by radar will be increased, alerting the defences earlier and, unless the altitude is extreme, the aircraft will find itself in the optimum engagement envelope of many long-range surface-to-air missiles. Nor does high altitude make for precision attack.

Medium altitude is often a reasonable compromise, offering good fuel economy and endurance while placing the aircraft above the effective reach of many surface-to-air systems, and while detection distance is still far too great for comfort medium altitude does permit accurate target location via ground mapping radar, especially using Doppler beam sharpening, and an accurate diving attack.

Low-altitude penetration also has both advantages and disadvantages:

it can sometimes avoid detection altogether, while exposure time is measured in seconds, often too few for defensive systems to track and fire. On the other hand, it compounds the difficulties of navigation and target location, accurate aiming becomes a problem, and without special equipment the mission cannot be flown at night or in adverse weather. Finally it is heavy on fuel and restricts range.

The accent is currently on low-level penetration, but medium and high level can be used in some circumstances: high-altitude, high-speed penetration is really the province of the strategic bomber armed with long-range stand-off missiles, but tactical penetration can be made where the defences are weak with the

Above right: The original A-10 mission would have involved a version of World War II cab-rank tactics, with loitering Thunderbolts ready to support ground forces as required.

Right: Revised tactics to deal with intense air defences emphasise sea-level penetration, smart weapons and ECM.

Breda Twin 40L70 aircraft engagement

As attack aircraft have been forced lower, air defences have adapted to the threat: with modern fire control and proximity-fuzed ammunition, Breda calculate that even a supersonic attack could expect to sustain hits by 3,578 splinters in this scenario, with two Twin 40L70 guns defending.

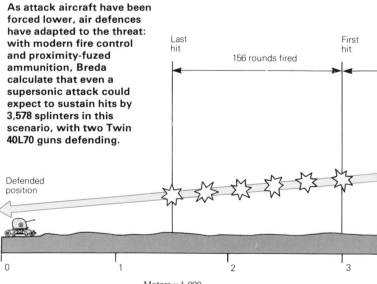

A-10 designed CAS mission

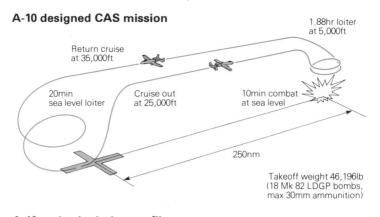

1.88hr loiter at 5,000ft

Return cruise at 35,000ft

20min sea level loiter

Cruise out at 25,000ft

10min combat at sea level

250nm

Takeoff weight 46,196lb
(18 Mk 82 LDGP bombs, max 30mm ammunition)

A-10 revised mission profile

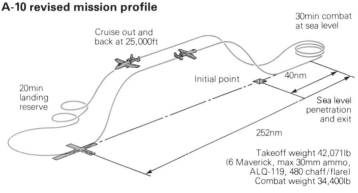

30min combat at sea level

Cruise out and back at 25,000ft

20min landing reserve

Initial point

40nm

Sea level penetration and exit

252nm

Takeoff weight 42,071lb
(6 Maverick, max 30mm ammo,
ALQ-119, 480 chaff/flare)
Combat weight 34,400lb

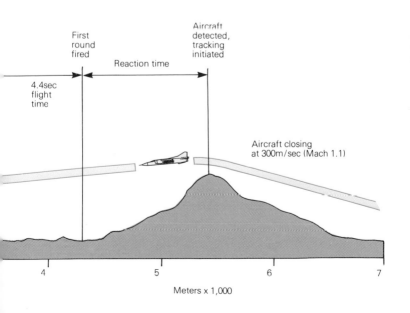

First round fired

Aircraft detected, tracking initiated

Reaction time

4.4sec flight time

Aircraft closing at 300m/sec (Mach 1.1)

4   5   6   7

Meters x 1,000

help of active and passive counter-measures and the weapons launched before the full strength of the defences is encountered. While this form of attack can be used against land targets, it can be particularly effective at sea — in a mass attack on a carrier task force, for example.

Medium-altitude penetration relies heavily on countermeasures and defence suppression, and while it is effective against moderate defences doubts are often expressed about its validity in a modern war zone. In 1983 a US carrier task force launched a strike on Lebanon and lost two aircraft, an Intruder and a Corsair. Given the current preoccupation with ultra low level penetration, ground defences are also concentrating on this area, possibly to the detriment of other defence levels, but it must be said that medium-altitude penetration against anything other than the weakest defences would require a great back-up effort, involving Wild Weasel defence suppression aircraft and specialised electronic warfare aircraft such as the EA-6 Prowler and EF-111 Raven, plus fighters carrying out sweeps and flying barrier patrols.

At ultra low level the attack aircraft flies in an invisible corridor with well defined boundaries set by altitude, speed and flight capability. The corridor tends to vary for different aircraft, depending on their gust response and whether they have specialised low-flying avionics. Low-flying systems are of the greatest importance: air defence systems vary in effectiveness, but the ground always rates 100%.

Where possible the mission profile will normally be hi-lo-lo-hi, with a high-altitude approach over friendly territory at an economical cruising speed followed by a descent to low level for a rapid subsonic penetration. The homeward leg will be the reverse, though if enemy fighters are encountered and fuel permits the low-level egress will be supersonic. Such a procedure can be damaging to the enemy in its own right: tests have shown that an aircraft at low level and supersonic speed sets up a shock wave than can damage sensitive electronic equipment, overturn soft vehicles and cause hearing impairment to troops.

Navigation is the next problem. The most basic method is to designate the target in a grid square and fly there by a direct route using map, compass, stop watch and a great deal of mental agility.

This method works in clear weather against unsophisticated defences where only a shallow penetration is needed, but it will be rare for all three conditions to be met, and modern attack fighters are equipped with nav/attack systems of varying degrees of sophistication. Ideally, the route should be carefully planned beforehand to delay detection for as long as possible, to avoid defen-

Tornado lo-lo-lo-lo mission

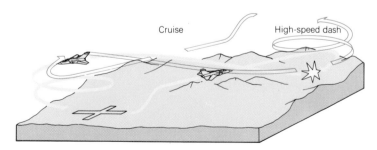

Cruise   High-speed dash

Above: The Tornado is capable of missions of 500nm or more radius with a typical weapons load without flying higher than 200ft. The bulk of the mission is flown at cruising speed, followed by a high-speed dash over the target: the point at which the dash will be initiated will be determined by estimates of the defences likely to be encountered.

Low-level strike corridor

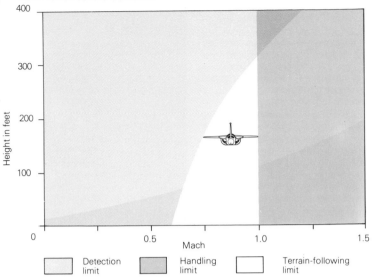

Detection limit	Handling limit	Terrain-following limit

sive strongpoints, to confuse the defenders as to intent and, finally, to give the best possible line of approach to the target combined with the most suitable heading for the bug-out.

The pilot's workload is high. He has to fly the aeroplane accurately at low level, keeping a sharp lookout for obstacles ahead, enemy fighters behind and SAMs anywhere while staying on course, identifying turning points and making adjustments to fly around patches of weather if he is not

Above: The parameters of the attack corridor will vary, but at low levels Mach 1 will be a practical limit, leaving a balance to be struck between keeping clear of terrain and staying hidden from radar.

equipped to go through them. This is where a good nav/attack system really comes into its own. At its most basic it will include an inertial navigation system (INS) preset to the known position on the ground before

Tornado hi-lo-lo-hi mission

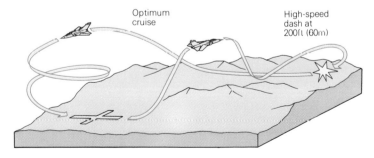

Optimum cruise

High-speed dash at 200ft (60m)

Above: If it is possible to fly the outward and return legs of the mission at optimum cruise altitude the Tornado's mission radius rises to more than 850nm, even allowing for the high-speed

penetration dash to and from the target which is unlikely to be less than 100nm. The weapons load is not specified, but would certainly include ECM pods as well as offensive stores.

take off and able to maintain a high degree of accuracy during the mission — no more than a couple of miles' error per hour.

A Doppler can add to the accuracy by feeding an accurate speed over the ground into the computer since ground speed differs from air speed as shown on the ASI according to wind speed and direction and barometric pressure variation. A moving map display as used in the Harrier gives a continually updated position over the ground; otherwise the information is generally given in terms of coordinates on either head-up or head-down displays. Waypoints can be stored in the computer, which tells the pilot when and how much to turn. A head-up display is invaluable, as when a pilot is flying manually at low level looking down inside the cockpit is to be avoided if possible, although head-down displays are generally located at the top of the dash, where they are within the pilot's peripheral vision.

War does not stop at sunset, and darkness and adverse visibility degrade optically laid counter-air weapons, while cloud, rain and fog reduce the effectiveness of infra-red homing missiles quite dramatically. That helps attack aircraft penetrate the defences unscathed, but it compounds the difficulties of navigating accurately to the target, and a more sophisticated avionics fit is needed, often with a second crew member to share the workload. The increased capacity of a two-man crew also allows a more capable countermeasures suite to be included.

The first priority is to allow the aircraft to fly the same low-level mission as it would in daylight, which means avoiding contact with the ground, and three main systems exist, with varying capabilities. Forward-looking infra red (flir) pierces the darkness and presents a TV-quality picture on a screen in the cockpit, allowing the pilot to fly much as he would in

F/A-18 radar attack

Above right: A prominent navaid in the Harrier's cockpit is the Ferranti moving map display.

Right: High-resolution radar mapping is the key to the Hornet's attack capability.

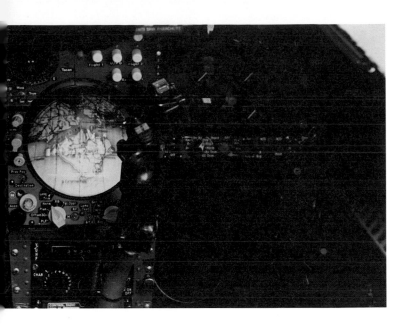

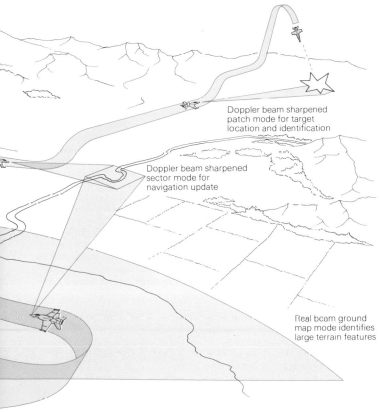

Doppler beam sharpened patch mode for target location and identification

Doppler beam sharpened sector mode for navigation update

Real beam ground map mode identifies large terrain features

137

Right: Hornet Flir image. Infra-red is most useful for targeting in the final stages of an attack.

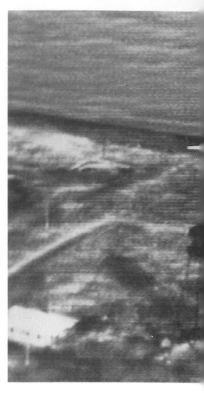

daylight, although of course the field of view (FOV) is far more limited. Flir gives a night capability, but its adverse weather capability is strictly limited, and as it is in service at the moment, Flir is more used for night attack than penetrating the target. Currently under development is Lantirn (Low-altitude navigation and targeting by infra-red at night), which will give a round-the-clock capability to the F-16, A-10 and other aircraft.

Next comes terrain-avoidance radar, whose elongated oval-section radar scan in front of the aircraft combines with a computer-generated template to warn of rising ground ahead and flash flightpath commands onto a screen. Various clearance levels can be preset — 250ft (76m), 500ft (152m) or 1,000ft (305m) are usual, or intermediate settings can be used if desirable. The system does not prevent the pilot from flying into the ground; it simply gives him the information necessary to avoid it.

Finally there is terrain-following radar, which, linked to the autopilot, actually flies the aircraft close to a preset height over the ground. Again, various clearance levels can be selected, normally between 200ft (60m) and 1,000ft (305m), and it is also possible for various grades of ride to be selected, normally hard, medium or soft. Hard ride keeps the aeroplane closest to the selected altitude, but at a considerable cost in crew comfort, and would normally only be selected to traverse heavily defended areas. Other settings are selected according to the threat. The transition to automatic terrain following mode does not mean that the pilot sits with nothing to do: apart from monitoring navigation functions, fuel states and so on he keeps a watchful eye on the TF radar presentation to ensure that it is working correctly.

On a deep penetration mission the navigation system needs to be updated at intervals to obtain the degree of accuracy necessary for a first-pass blind strike to be made. On the flight plan, which is often stored on a cassette tape and fed into the com-

Tornado terrain following

Terrain-following radar generates a theoretical ski-toe shaped envelope projected forward of the aircraft, and compares this with the profile of the terrain ahead. In the case of Tornado, penetration of the envelope by the terrain generates an automatic climb command which is passed to the autopilot and flight-director computers, resulting in an input to the control surfaces, and to the pilot's head-up and head-down displays, as shown here.

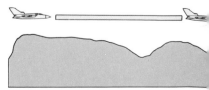

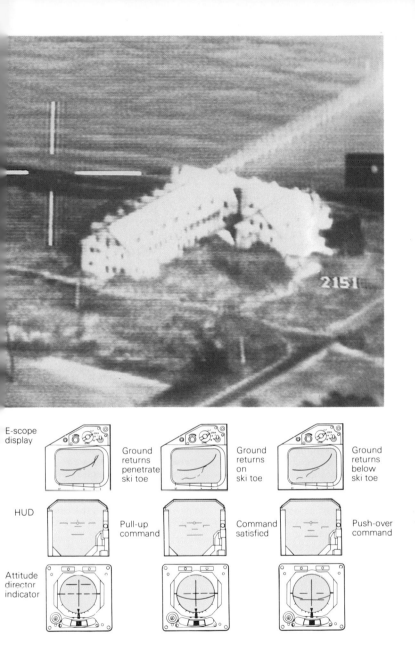

E-scope display			
	Ground returns penetrate ski toe	Ground returns on ski toe	Ground returns below ski toe
HUD	Pull-up command	Command satisfied	Push-over command
Attitude director indicator			

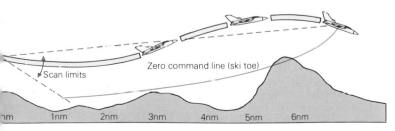

Scan limits   Zero command line (ski toe)

nm   1nm   2nm   3nm   4nm   5nm   6nm

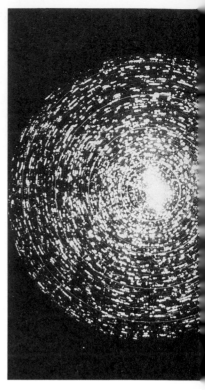

puter in the cockpit, various radar-reflective or radar-significant points will have been stored, and as these points are approached — not necessarily too closely — a few sweeps are made with the radar and their exact position established. This information is fed into the computer, which then updates the actual aircraft position, giving a very high degree of accuracy.

The final avionics aids to penetration are countermeasures, which are both active and passive. Radar warning receivers (RWRs) of varying degrees of complexity are becoming standard on attack aircraft: they detect when the aircraft is being painted by radar, and at the most basic level give an aural warning with a visual indication of which quadrant the hostile emission is coming from. At the top end of the market they not only detect a multiplicity of emissions but also classify them according to their nature — search, SAM, air-to-air or whatever — with a fairly accurate assessment of bearing, range, function and whether or not they constitute a threat, all of which is displayed in the cockpit, though in very intensive areas only those considered to be the greatest threat will be indicated. RWRs may even be able to take direct action in the form of jamming without the intervention of a crew member, although they can always be overridden at need. A typical example would be a low and fast aircraft acquired by radar in a position where it would be able to shake off detection by using terrain masking, in which case active jamming, being an emission, might continue to betray its position.

Jamming can consist of expendables in the form of flares or chaff, or active noise or deception jamming. ECM pods can be carried, although pods sterilise pylons that could otherwise be used for something more offensive, or an ECM suite can be fitted internally. ECM suites tend to be very comprehensive, and are designed to be flexible in countering new threats through programmability. Software

Inverse gain deception jamming

Deception jamming involves the transmission of fake return signals which the victim will accept as genuine. These are

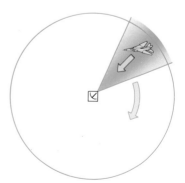

Normal operation: the antenna is pointing at the target aircraft, which returns a genuine echo seen on the PPI (left) as a target

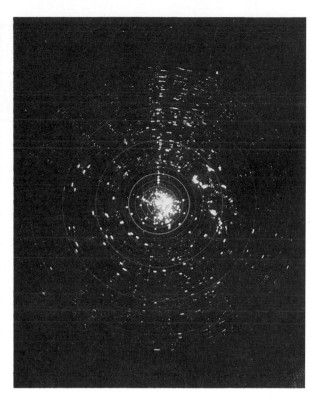

often sent when the antenna of the victim is not directly facing the jammer, so they must be very powerful in order to leak into the antenna via a sidelobe. The fake pulse will then be of an appropriate strength to be accepted as genuine.

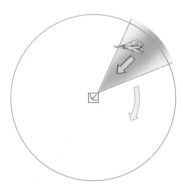

With the aircraft on the edge of the beam, the radar will reject the true target in favour of a powerful fake echo apparently on another bearing

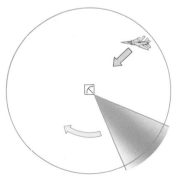

By transmitting a massive fake echo later in the scan pattern, the jammer can create another target on a totally false bearing

can be altered to meet a changing threat far more easily and cheaply than hardware, the main difficulty being deciding at what command level the authority to initiate a change should be vested, which is of course a wartime problem; in peacetime speed is not of the essence.

Having successfully penetrated the air defence system, it is vital for the attacker to locate the target in time for a first-pass attack; second time around is simply not good enough, as an abortive first pass will only alert the defenders. Another factor is fuel; in many cases there will be insufficient for feints or other deceptive measures if the first pass fails.

At the most basic level, target acquisition will be visual, while successive steps of sophistication lead up through electro-optical means such as flir or televisual acquisition, ground-mapping radar and offset blind bombing to the use of synthetic aperture radar. For all practical purposes there are two main types of targets, those in known locations and those that are mobile; while the latter may be expected to be in a certain area, their exact location is not known. There will also be occasions when previously undetected targets will be discovered.

Visual detection is dependent on daylight and clear weather. Even then a fast jet at low level stands little chance of visually acquiring anything other than a large area target unless it appears fortuitously straight ahead. Normal visual acquisition depends on

the aircraft leaving the shelter of the ground briefly for a pop-up to have a quick look round. The tactical wisdom of such a manoeuvre will depend on the strength of the defences in the specific area; in some cases it will be feasible, in others suicidal.

Flir presents a picture on the screen made up from heat imagery; ideally this is displayed at the exact size that the pilot would see visually, while the focus can be adjusted if required to give a close-up view. The picture is in black and white, and definition in clear air is very good, turning night into day, although only along the line of sight. While generally reckoned to give a night attack capability, flir can also be used to penetrate smoke and dust over the battlefield.

Televisual aids can also be used, as with Maverick; the camera in the nose of the missile displays a picture in the cockpit which the pilot uses to lock on the missile using brightness contrast (light against dark or vice versa). Unfortunately, a high level of contrast is needed, which calls for clear weather and means the direction of attack needs to take the angle of the sun into account to give the greatest amount of contrast. That is why the latest version of Maverick uses IIR for homing and contrast.

Ground-mapping radar is used to scan the terrain ahead and present a picture in the cockpit from which

Below: An airfield target several miles ahead as shown on the radar display of a Tornado.

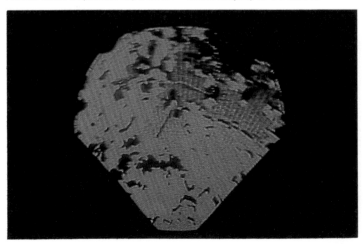

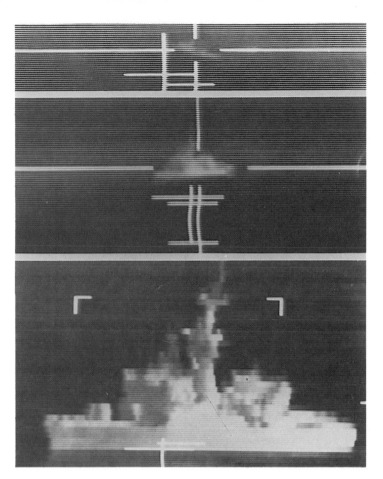

Below: At closer range the target is marked using the hand controller ready for an attack.

Above: Maverick IR image of a US destroyer at acquisition, visual and terminal ranges.

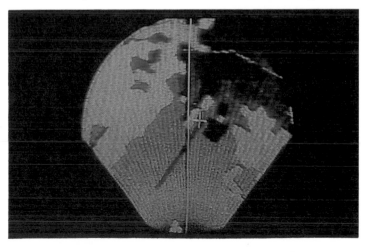

targets may be identified. As slant look angles cause a fair amount of distortion, a computer is used to process the returns and present the picture in plan view, so that it is more easily identified.

Blind bombing is carried out by approaching an easily identifiable point, visually or by radar, at a precomputed speed, altitude, and heading. Once there, the aircraft is pulled into a climb, typically of 30°, and the bombs are released, after which they travel on under their own momentum for about three miles (5km), reaching an altitude of about 3,000ft (914m) en route. This method allows full use of terrain masking to be made. Once the bombs have been released the aircraft is free to reverse course and return to low level.

If no suitable identifiable point exists close to the target an offset point can be used, with a radar return from it fed into the attack computer to give a very accurate position for the aircraft. This form of bombing is exploited even more by using the Rockwell GBU-15, which can be launched as described, the target being acquired via either a television camera or an IIR seeker while the weapon is in flight and lock-on being made by data link.

Synthetic aperture radar (SAR) was first developed for reconnaissance purposes back in the 1960s, but since then advances in processing have allowed the definition to be improved to the level of low-grade photography, an improvement not only good enough to give targeting data, but also one which can be used to detect, and often identify, previously unknown targets to one side or the other of the flight path.

Guided weapons excepted, the accurate release of air-to-ground munitions is beset with difficulties. Bomb or rocket ballistics, aircraft speed, which may reach 900ft (275m) per second, aircraft velocity vector, weapon release parameters, range and even cross-winds all affect the aiming equation, which with modern equipment is taken care of by the stores management and weapons aiming systems. Obviously, the precise moment of weapon release is critical, and a split second delay on

the part of the human operator could cause a gross error, so with many weapons release is also automated.

Irrespective of whether the attack mode is lay-down, toss bombing, low angle dive bombing or dive-toss, the

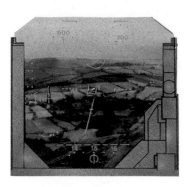

Above: Tornado HUD symbology for a straight-pass attack using an offset aiming point in automatic terrain-following mode.

Below: Offset aiming is used where a target is unlikely to show up clearly on radar. When planning the sortie the crew select a nearby radar-prominent feature — in this case a pylon — and load its coordinates into the nav/attack system along with those of the target.

Offset bomb-aiming

Synthetic aperture radar

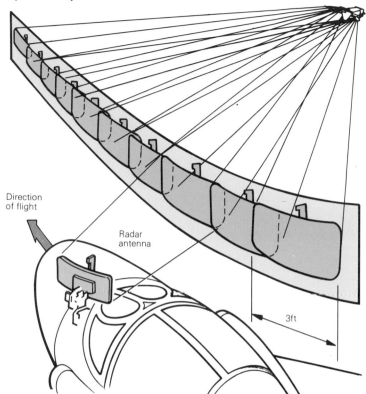

Direction of flight

Radar antenna

3ft

Above: Synthetic aperture radar involves collating a series of returns and analysing differing return angles on left and right halves of the antenna to give a target's relative position.

data is fed into the computer beforehand, any time between the mission planning stage and the approach, depending on the weapons and the target. As the attack is commenced the bomb release button is depressed, a process known as pickling, and the bombs will be released when the electrons think they have solved the problem, or when a preset range has been reached. The crew's task is thus simplified; having fed the required data into the number cruncher, all they have to do is to arm the system at the right time, by means of the pickle button, then fly the aircraft accurately towards or over the target.

Two acronyms that occur often in the weapons aiming context are CCIP and CCRP, along with their derivatives CCIP/IP and CCRP/IP.

CCIP stands for continuous computation of impact point, sometimes irreverently known as the death dot, while CCRP is continuous computation of release point. IP stands for initial point.

When flying low over the battlefield or the target area, a weapon released at a given moment will impact at a given spot. The CCIP is that spot, and can be switched on ready for the possibility of a target appearing under the aircraft's flightpath. This instantly available aim point makes a preselected weapon immediately ready for use. With CCIP/IP, the coordinates of a known

Right: An F/A-18 lets go a pair of Mk 82 slicks in a medium-altitude shallow diving attack.

Direct delivery bombing

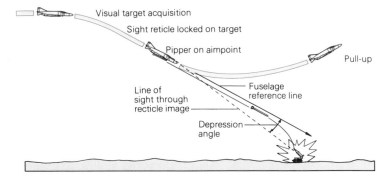

Above: Direct delivery mode is used only as a last resort when a faulty INS has fed incorrect data to the weapons release computer; weapons are released on pickle.

Below: Dive-toss mode gives improved accuracy and allows for evasive manoeuvres during the approach: with the pipper on target, release is automatic.

Dive-toss bombing

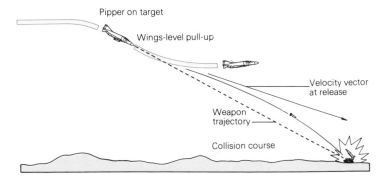

Dive-glide bombing

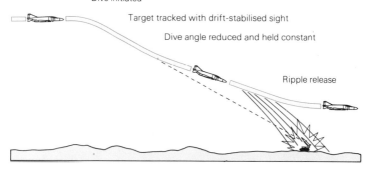

Dive initiated

Target tracked with drift-stabilised sight

Dive angle reduced and held constant

Ripple release

Above: Dive-glide release is used against area targets. The pilot hits the pickle button when the pipper is on target, and the computer carries out the release.

Below: Dive-level or lay-down bombing is used with low-drag weapons and involves the pilot maintaining a constant ground track through the target.

Dive-level bombing

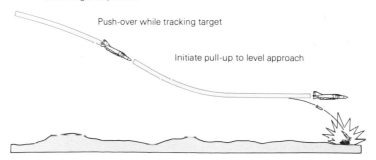

Visual target acquisition

Push-over while tracking target

Initiate pull-up to level approach

point — the initial point — are stored in the nav/attack system before takeoff, together with its relationship to the target: once the IP is reached and designated piloting information is displayed, probably on the HUD, which enables the pilot to fly accurately to the target, and the weapon is released automatically at the appropriate moment. The CCIP is adjusted automatically for height, speed and weapon selected, including guns.

CCRP is similar but is used for toss of loft bombing; once the target has been designated weapon release is automatic. CCRP/IP also incorporates an IP which works in exactly the same way as the CCIP/IP: piloting instructions are given in similar fashion and weapons release is again automatic. The degree of automation in both CCIP and CCRP removes one potential cause of aiming error.

The missions that the attack aircraft will be called on to fly vary between close air support and long range interdiction, or possibly anti-shipping sorties. All have their various requirements and tactical approaches.

Close air support/battlefield air interdiction involves shallower penetration of hostile airspace than any other mission. In this scenario targets are frequently in close proximity to friendly forces, and identification is made difficult by smoke and dust, without the additional complication of poor visibility. For conventional fast jets, including armed trainers, target acquisition and identification becomes a major problem. Ideally they would be assisted by a forward air controller (FAC), based either on the ground or in a helicopter, who would allocate targets and direct attacks. The FAC might be in touch with, or even directing, a laser designation team, which would ease the difficulties somewhat; the target would be picked up by the laser kit in the nose of the aircraft and attacked without even being seen. Ideally the attack run would be made from behind the FLOT, the weapons delivered and a prompt egress initiated. Aircraft would operate in two-ship elements as a basic fighting formation, and with luck many friendly aircraft would be in the area at one time to confuse and saturate the defences. The first priority in this mission would be to knock out hostile ground-based air defence systems, mobile radars and AA guns; the CAS aircraft could then set about enemy armour and APCs.

The A-10 Thunderbolt is the only modern aircraft known to have been designed as a tank-killer. One of the difficulties of air action close to the FLOT is that the enemy forces are deployed, presenting widely spaced targets which are not very suitable for attack by anything other than precision weapons. A fast jet stands little chance of knocking out a single tank unless its flight path takes it almost straight overhead, and even then there is little time to select a weapon and aim. The A-10, by contrast, may be classed as a slow jet, relying on armour and redundant systems to survive hits rather than speed to avoid hits in the first place. Its relatively low speed confers many benefits, reducing the radius of turn to allow the aircraft to remain over one portion of the battlefield with ease and stay below a low cloud base, giving the pilot more time for positive identification and permitting

Pave Penny attack scenario

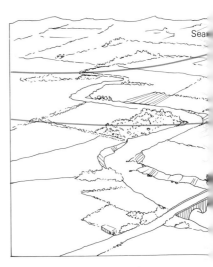

Above: Ground-based laser designation can be of great help to a pilot carrying out a close-support mission. This is the Ferranti Battlefield Operations Laser Designator used by British troops to designate targets for laser-guided bombs during the Falklands War of 1982.

Below: The AAS-38 Pave Penny pod, standard equipment on the A-10, is able to acquire targets designated by laser-equipped forward air controllers. The pod detects the reflected radiation and indicates the target on the pilot's HUD, enabling the pilot to manoeuvre for an attack.

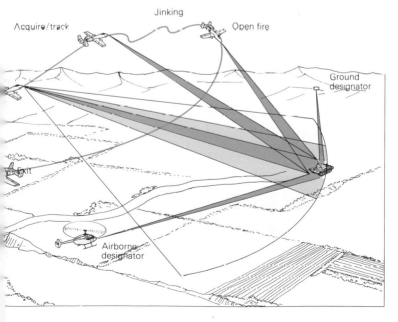

him to fly really close to the ground and use terrain masking.

The A-10 carries precision weapons — AGM-65 Maverick, which is a launch-and-leave weapon, and the giant GAU-8/A Avenger gun, with its depleted uranium-cored shells. Approaching from behind friendly lines at low level, it pops up briefly to around 700ft (210m), acquires a target, locks on a Maverick, launches and dives away, turning as it does so. The same procedure is followed with the gun, except that the range is rather shorter. The one thing the A-10 is not intended to do is penetrate hostile airspace; if it goes more than 2nm (3.75km) over, it has gone too far.

Behind the battlefield is the interdiction zone. A modern battle is heavily dependent on fuel, munitions and reinforcements, and the battle zone can be roughly defined by the range of modern artillery — within about 15nm (28km) of hostile positions land forces can reasonably be assumed to be deployed, and it is behind this area that the juiciest targets are to be found. Speed of movement, as reinforcements are hurried forward, will dictate that

A-10 low-level gun attack

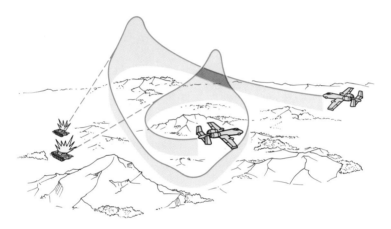

Above: The A-10 spends most of its time less than 100ft (30m) above the ground, popping up briefly to a maximum of 500ft to deliver short bursts of fire.

Below: Terrain masking and three-dimensional evasive jinking are also employed in the run-up to a Maverick launch from 500ft.

A-10 low-level Maverick delivery

roads are heavily used; roads have to cross rivers and wind through defiles, which constitute choke points. Further back railways will be attacked, and particularly marshalling yards. Once a target is located, it will be hit in force by a dozen or more aircraft, probably using area weapons such as cluster bombs. Harriers, with their forward basing and rapid reaction times, would be used against area targets just behind the deployment zone, while Jaguars, F-16s and Mirages hit targets further back. Still deeper in, Tornados and F-111s would strike at communications,

although they are at first more likely to fight the counter-air battle by strikes against airfields.

If there is any choice, the deep penetration missions will be flown at night or in adverse weather to hamper the defences. Airfields are heavily defended targets and the use of laydown weapons, which involve overflight, has been heavily criticised in some quarters, but while stand-off weapons may be preferable from some viewpoints, the attacking aircraft has to pull up to acquire the target and lock on before release, which makes it vulnerable. An alter-

ZSU-23-4 avoidance tactics

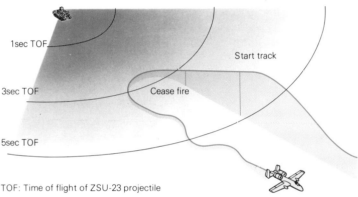

3sec linear flight path from start track to cease fire; open fire after 1.5sec

1sec TOF

3sec TOF

5sec TOF

Start track

Cease fire

TOF: Time of flight of ZSU-23 projectile

Above: At GAU-8/A range an A-10 can deliver a 1.5sec burst and be back under a Shilka's minimum elevation before the 23mm projectiles can reach it.

Below: The A-10 should be able to return to terrain masking after the attack in less time than it takes the SA-8 to acquire, lock on and launch.

SA-8 avoidance tactics

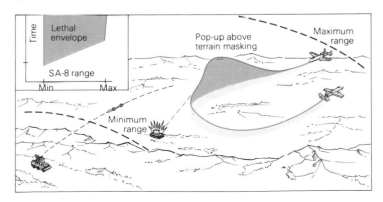

Lethal envelope

Time

SA-8 range

Min   Max

Pop-up above terrain masking

Maximum range

Minimum range

native view is that if a minimum of eight aircraft are sent to attack an airfield, with two or four toss-bombing with conventional bombs set to airburst in order to keep the defenders' heads down while the others run in from different directions and in rapid succession with laydown weapons, casualties should be light.

Successful attacks on bridges and other small, hard targets demand pinpoint accuracy. Night attacks are preferred, not only for defensive purposes but also because the majority of enemy movement will take place at night: a bridge dropped early in the evening will cause far more of a bottleneck than one knocked down first thing in the morning, as temporary repairs will be more difficult, the logistics timetable will be more disrupted, and the back-up of supply vehicles should provide a rich target through the following day.

LGBs are favoured for this type of attack, delivered, for example, by F-111Fs equipped with Pave Tack designator pods. The approach would be flown fast and low to an IP, where the nav/attack system would be updated, and at a fixed distance from the target the aircraft would pop up to acquire it on radar and release the bombs using radar aiming. Immediately after release, and while the aircraft is turning away and back to low level, the WSO in the right hand seat acquires the target on Pave Tack, which has an infra-red sensor, then switches on the laser designator to track it: the bombs should then acquire the laser reflections and home on them. The process sounds very simple, but in fact it is very difficult and requires a long period of intense training. The pilot has to fly a carefully calculated manoeuvre manually in darkness, while the WSO has to operate Pave Tack while being flung around the sky, sometimes half inverted, concentrating solely on holding the designator on target.

Defence suppression is an inevitable part of any deep penetration mission. While aircraft like Tornado and many others can carry antiradiation missiles, the USAF has a specialised defence suppression aircraft, the F-4G Wild Weasel, fitted with an electronic system which can detect and classify hostile radar emis-

F-111 Pave Tack deployment

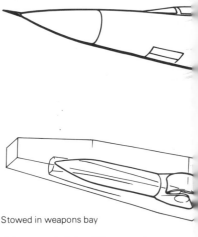

Stowed in weapons bay

Above and right: When not in use the Pave Tack pod is stowed in an F-111F's weapons bay. On the approach to the target it is lowered and rotated and pointed within the limits shown.

Below: Pave Tack combines a Flir viewing system which enables the Weapon Systems Officer to designate a target with a laser ranger/ designator for accuracy.

Pave Tack operation

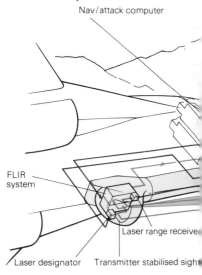

Nav/attack computer

FLIR system

Laser range receiver

Laser designator   Transmitter stabilised sight

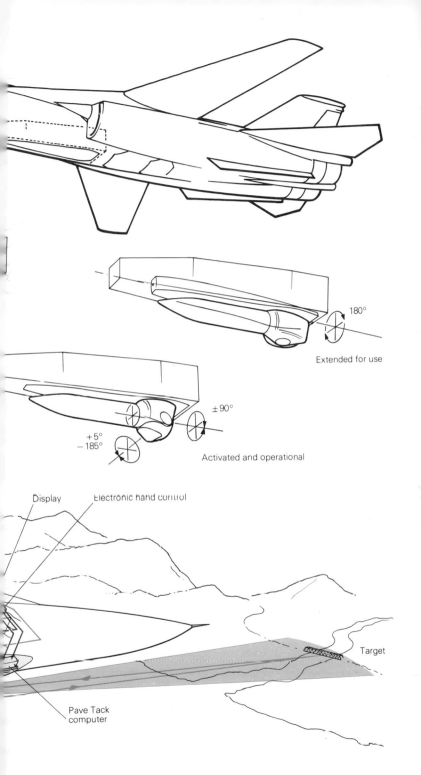

180°

Extended for use

±90°

+5°
−185°

Activated and operational

Display   Electronic hand control

Target

Pave Tack
computer

153

sions; with a minimum of three bearings, it can pinpoint missile radars and attack them. It flies low, occasionally popping up above the radar horizon to receive emissions and take bearings. The types and locations of radars are collected and stored, and when sufficient is known about the defences in a certain area, the attack begins. Wild Weasels operate in pairs, an F-4G teamed with an F-4E, whose function is only to attack; this would be co-ordinated from different directions, using anti-radiation missiles and CBUs. Ideally, the team would consist of two F-4Gs, but they are expensive, and there are not enough of them to go round.

F-4Gs would be used to support strike forces, defence suppression not being an end in itself. Weasel effectiveness may well be improved in the future by the use of Situational Awareness Technology, or SAT, which is currently under development by LTV Aerospace. This teams three F-16s, equipped with a SAT display fed by data link, with a single F-4G. The F-4G thus becomes a hunter with a killer F-16 on his wing, plus a pair of F-16 killers in attendance. The detection data gathered by the F-4G is passed to the F-16s, which launch co-ordinated attacks on the targets.

Finally, there is the anti-ship mission. This is accomplished mainly by stealth, with missile-carrying aircraft flying out to the search area at low level before climbing to a few thousand feet for a quick radar scan. If a target is found the coordinates are fed into the navigation system of the missile, which is then launched, flying most of the way at low level on a preprogrammed course before switching to on-board homing, usually active radar, for the final leg. It is possible for the detecting aircraft to pass the data onto a companion at a considerable distance, and still at low level, who will then launch a missile from a position undetectable by the surface ships. This is essentially a medium- to long-range form of attack, and the requirements of the aircraft carrying it out are simply an adequate radar and the ability to carry the missile a reasonable distance and launch it. Of course, if the target happened to be an American carrier task

group defended by the Hawkeye/Tomcat combination, it would also need a lot of luck.

Right: With four Kormoran, Sea Eagle or Harpoon missiles, Tornado has a range in the maritime strike role of better than 700nm from base to the weapon's stand-off range. Optimum altitude cruise out and back is combined with a high-speed dash to the release point.

Right: Harpoon is launched at a typical stand-off range of 50nm and at medium altitude. It is programmed to dive to low level for the inertially guided mid-course phase, then to sea-skimming height for the active radar-guided terminal phase.

Below: Following a similar attack profile, Kormoran hits just above the waterline.

Tornado maritime strike

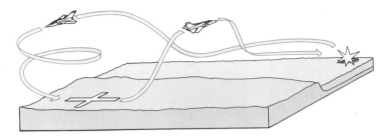

Harpoon attack

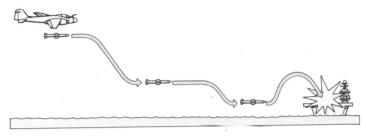

OTHER SUPER-VALUE MILITARY GUIDES IN THIS SERIES......

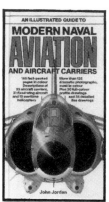

Air War over Vietnam
Allied Fighters of World War II
Battleships and Battlecruisers
Bombers of World War II
Electronic Warfare
German, Italian and Japanese Fighters
  of World War II
Israeli Air Force
Military Helicopters
Modern Destroyers
Modern Elite Forces
Modern Fighters and Attack Aircraft
Modern Soviet Air Force
Modern Soviet Ground Forces

Modern Soviet Navy
Modern Sub Hunters
Modern Submarines
Modern Tanks
Modern US Air Force
Modern US Army
Modern US Navy
Modern Warships
NATO Fighters
Pistols and Revolvers
Rifles and Sub-machine Guns
Space Warfare
World War II Tanks

＊Each has 160 fact-filled pages
＊Each is colourfully illustrated with hundreds of action photos and technical drawings
＊Each contains concisely presented data and accurate descriptions of major international weapons
＊Each represents tremendous value

If you would like further information on any of our titles please write to:
Publicity Dept. (Military Div.), Salamander Books Ltd.,
52 Bedford Row, London WC1R 4LR

PRINTED IN BELGIUM BY
proost
INTERNATIONAL BOOK PRODUCTION